SUPERNOVA

The MIT Press Essential Knowledge Series

A complete list of the titles in this series appears at the back of this book.

SUPERNOVA

OR GRAUR

The MIT Press | Cambridge, Massachusetts | London, England

The MIT Press would like to thank the anonymous peer reviewers who provided comments on drafts of this book. The generous work of academic experts is essential for establishing the authority and quality of our publications. We acknowledge with gratitude the contributions of these otherwise uncredited readers.

This book was set in Chaparral Pro by New Best-set Typesetters Ltd. Printed and bound in the United States of America.

Library of Congress Cataloging-in-Publication Data

Names: Graur, Or, author.
Title: Supernova / Or Graur.
Other titles: MIT Press essential knowledge series.
Description: Cambridge, Massachusetts : The MIT Press, [2022] | Series:
    MIT Press essential knowledge series | Includes bibliographical references
    and index.
Identifiers: LCCN 2021000487 | ISBN 9780262543149 (paperback)
Subjects: LCSH: Supernovae.
Classification: LCC QB843.S95 G738 2022 | DDC 523.8/4465—dc23
LC record available at https://lccn.loc.gov/2021000487

10   9   8   7   6   5   4   3   2   1

Dedicated to BUF19Sop (SN Sophia) and BUF19Awd (SN Awdrey).

CONTENTS

Series Foreword   ix
Introduction   xi

1   Supernovae through the Ages   1
2   Supernovae in the Modern Era   19
3   Enlightened by Light   39
4   The Lives and Deaths of Stars   57
5   Supernovae, the Universe, and Us   81
6   Supernovae as Tools   99
7   Burning Questions   121
8   A Bright Future   143
    Conclusions   167

    Acknowledgments   171
    Units and Elementary Particles   173
    Glossary   175
    Notes   183
    Further Reading   203
    Index   205

SERIES FOREWORD

The MIT Press Essential Knowledge series offers accessible, concise, beautifully produced pocket-size books on topics of current interest. Written by leading thinkers, the books in this series deliver expert overviews of subjects that range from the cultural and the historical to the scientific and the technical.

In today's era of instant information gratification, we have ready access to opinions, rationalizations, and superficial descriptions. Much harder to come by is the foundational knowledge that informs a principled understanding of the world. Essential Knowledge books fill that need. Synthesizing specialized subject matter for nonspecialists and engaging critical topics through fundamentals, each of these compact volumes offers readers a point of access to complex ideas.

"I've never seen a supernova blow up, but if it's anything like my old Chevy Nova, it'll light up the sky!"
—Philip J. Fry[1]

Sunlight streams through the tall glass windows of the *Daily Planet* offices where mild-mannered reporter Clark Kent, about to be fired by his editor, leaps out. Sapped of his superhuman strength, the man of steel plummets through the air. Seconds later, he is saved by a mysterious new superhero that has shown up in Metropolis, a masked hero called Supernova! Who is the hero behind the mask? Who is this Supernova, whose radiance momentarily outshines the Sun?[2]

Meanwhile, in a small gallery in London, Sherlock Holmes frantically pores over the canvas of a lost Vermeer painting. He has ten seconds to prove the painting is a fake; the life of a young boy, explosives strapped to his body, hangs in the balance. At the very last moment, Sherlock cries: "The Van Buren Supernova! A huge star blowing up. Only appeared in the sky in 1858!" As Inspector Lestrade looks dumbfounded at the amateur detective, Dr. John Watson peers at one of the small white blobs in the painted night sky and explains, "So how could it have been painted in the 1640s?"[3]

Halfway across the Galaxy, the crew of the USS *Voyager* witnesses a supernova explosion. The next day, two more supernovae go off in *Voyager*'s vicinity. On the bridge, Captain Kathryn Janeway turns to face Q, a seemingly omnipotent alien being who has appeared on her ship, and accuses him: "A star going supernova is an event that occurs once every century in this galaxy. Now we're about to witness our third in less than three days, all in the same sector. Why do I suspect you have something to do with this?" Captain Janeway, a trained scientist, is right. Even though the average frequency of supernova explosions in the Milky Way Galaxy is 2–4 supernovae per century, it is extremely unlikely for three of them to occur one after the other in the same small patch of space. The captain's day is about to get worse when she learns that the supernovae now buffeting her ship are caused by a brutal civil war raging in the Q Continuum.[4]

Fourteen years later, without any apparent intervention from Q, the star at the center of the Romulan star system explodes in what becomes known as the Romulan Supernova. It obliterates the Romulan home world of Romulus, kills tens of millions of Romulans, and launches a series of events that will see the destruction of Vulcan, the burning of Mars, and the resignation from Starfleet of a disillusioned Admiral Jean-Luc Picard.[5]

Based loosely on actual or purely imaginary astrophysical phenomena, these works of fiction allude to some of the most pressing questions facing astrophysicists today: What are supernovae? How can we use them to further our understanding of physics? And what effects do they have on the Universe and us?

Supernovae are the explosions of stars. These explosions are some of the most energetic phenomena in the Universe, rivaling the combined light of billions of stars and visible clear across the Universe. Plate 1 shows an example of a young supernova in a nearby galaxy; although several days past its peak, it is still one of the brightest objects in its host galaxy.

Just as we see different types of stars in the night sky, so there are different types of supernovae. Like us, stars are born, they mature, and eventually they die. How they live and die mostly depends on the mass they were born with,[6] but also on where they were born: the type and age of their host galaxy, as well as their specific neighborhood within that galaxy. A star's fate also depends on its marital status, so to speak: is it a loner, like our Sun, part of a couple living in a binary system, or a member of a more complicated relationship with two or more other stars? Over the last eighty years, astrophysicists have managed to connect some types of supernovae to specific types of stars, but there are still many connections left to make.

These connections between the supernovae we observe and their progenitor star systems give us an insight not only into the supernovae themselves but also into the

Supernovae are the explosions of stars. These explosions are some of the most energetic phenomena in the Universe, rivaling the combined light of billions of stars and visible clear across the Universe.

evolution of the stars, how they go from dispersed clouds of hydrogen, to fiery balls of burning gas, to hellish infernos of destruction. This, in turn, impacts our understanding of other fields in astrophysics and their own open questions: How are stars formed in different types of galaxies? How are cosmic rays (charged particles like protons and electrons, which regularly impact our atmosphere) accelerated to near-light speeds? And how are the various elements in the periodic table created?

A deep observational understanding of supernovae—why and how they shine, and how their brightness changes over time—also allows us to use them as tools for experiments in astrophysics as well as basic physics. A certain type of supernova, for example, brightens and fades in such a predictable manner that we can calibrate the explosions we observe and use them to measure the distances to their host galaxies. Two such experiments in the late 1990s led to the discovery that, instead of decelerating due to gravity as expected from general relativity, the expansion of the Universe was actually accelerating. Either gravity works differently than expected on cosmic scales or we need to take into account a new type of physical substance, dubbed "dark energy," which exerts a negative gravitational pressure.

Supernovae have also left their mark on Earth—and on us. Many of the elements that make up our bodies, including the iron in our blood, are created and expelled into space by supernova explosions. Our Sun is not massive enough to die as a supernova, but there is some evidence that a nearby supernova (or more than one) may have caused a massive extinction event on Earth some 2.6 million years ago during the Pliocene-Pleistocene boundary period. There is nothing to worry about, though. Although it is quite likely, in astronomical terms, that the Solar System will be affected by multiple supernovae in the next few billions of years, the likelihood that such an event would happen in our lifetimes (or those of our descendants for many generations to come) is vanishingly small.

As in any specialized field, jargon and technical details are often used to put up barriers between the field's practitioners and the uninitiated. This is a human tendency that has nothing to do with the "difficulty" of the field. In this book, I have tried to make the descriptions of our knowledge of supernovae as accessible as possible to all readers. Any jargon I felt was important to keep is explained when it first appears (highlighted in boldface) and is collected in the glossary at the end of the book. Unessential jargon, as well as technical details that might be of interest to other astrophysicists, can be found in the endnotes of every chapter. Two tables at the end of the book also list the names, symbols, and meanings of units and elementary particles used throughout the text.

A few notes about terminology and pronunciation. The word "supernova" derives from Tycho Brahe's Latin term "stella nova" (new star). As such, its plural is "supernovae." Although "supernovas" is sometimes used in popular media, it is seldom used by astronomers. The pronunciation of "supernovae" varies widely, depending on each person's pronunciation of Latin words, but most astronomers have come to use an "ee" sound at the end of the word. In the scientific literature, "supernova" will often be shortened to "SN" and the plural "supernovae" to "SNe." Supernovae are usually named after the year and order in which they were first observed. SN 1987A, for example, was the first supernova observed in 1987.

The terms **luminosity** and **brightness** are both used throughout this book. Luminosity is used to describe the amount of light output by a star or other astrophysical object, while brightness is used to describe how much light we receive from that object here on Earth. Thus, two stars may have the same luminosity, but the star that is closer to the Solar System will be brighter than the one that is farther away. Luminosity is objective, brightness subjective.

The Romans called the Mediterranean Sea Mare Nostrum: "Our Sea." In a similar vein, astronomers use capitalization to distinguish between astrophysical phenomena in general and the ones that are "ours." We orbit a star we call the Sun. There are many moons orbiting the planets of the Solar System, but ours is the only Moon. Our Solar System is one of many (though all others are called stellar systems) in the Milky Way Galaxy, which is just one of billions of galaxies in our particular Universe.

Throughout this book, I use "astronomers" and "astrophysicists" interchangeably. In the English language, the word "astronomer" has existed at least since the fourteenth century CE to denote someone who studies the movements of the stars and planets.[7] "Astrophysicist" came into use only in the late nineteenth century.[8] At first, astrophysicists were only those astronomers or physicists who used the then-new technique of spectroscopy (using a prism to scatter the light from an object into its constituent wavelengths) to study the physics and chemical makeup of the Sun and other stars.[9] By the turn of the twentieth century, "astrophysicist" had come to encompass any scientist seeking "to ascertain the nature of the heavenly bodies, rather than their positions or motions in space—what they are, rather than where they are."[10]

This definition has more or less endured for the last hundred years, but I would argue that nowadays, with few exceptions, all astronomers are astrophysicists. During the twentieth century, astronomy, physics, and chemistry evolved to the point where astronomy can no longer stand on its own without the other physical sciences. Today, I would argue that all astrophysicists are, at their core, physicists, and the mission statement of our field is: Use physics to understand the workings of the Universe, and use the Universe to understand physics.

This new mission statement is embedded in the way we study supernovae. To understand the nature of supernovae, we use physics (and chemistry) to analyze observations acquired through carefully designed experiments. At the same time, we use supernovae as tools in experiments in other fields of physics and astronomy.

For too long, popular culture has focused on a handful of famous, eccentric, or controversial scientists. Worse, these specific individuals tend to be white men of European descent. This has created a lasting but fundamentally wrong impression that science is done by a handful of white, male, socially challenged geniuses. In reality, there are tens of thousands of scientists spread across the world. The vast majority of us lead pretty normal lives. We raise families, engage in hobbies, and would never, ever consider ourselves geniuses. To combat this pernicious stereotype, I have sought to highlight the global and collaborative nature of astronomy and refrained from gossiping about the astrophysicists mentioned throughout the book. However, where the lives of these scientists have already been chronicled by others, I have included references to their biographies.

With this short book, I hope to provide a concise introduction to the study and use of supernovae, one of the more exciting fields of astrophysics and my own chosen specialty. That means that I have had to leave out quite a bit of fascinating observations and theory; I hope that none of my colleagues will be affronted if their work has been left out.

Every year, dozens of new papers are published in our field. Most of them report incremental advances in our understanding; truly ground-breaking discoveries or new ideas are rare. So, while some of the details presented in the coming chapters will surely be out of date in a few years, I expect that the pillars that support our understanding—the essential knowledge presented throughout this book—will hold. If I am proved wrong, that would mean that brand-new physics has been discovered, and the next edition of this book will be even more exciting to write.

SUPERNOVAE THROUGH THE AGES

Astronomy is often referred to as the oldest of the sciences. This is somewhat anachronistic, since our definition of science is barely two centuries old. It would be more accurate to describe astronomy as a confused but lively conversation stretching back for thousands of years. The names by which astronomers are called have changed over time: elders, shamans, priests; augurs, philosophers, astrologers; scientists. The languages they used and the reasons that prompted their observations have changed as well: describing the origins of the world, predicting the future, twitching the veil on the workings of the Universe. In one name or another, supernovae have been studied for more than 1,800 years, and modern studies are still very much shaped by those conducted hundreds of years ago.

Astronomy is a confused but lively conversation stretching back for thousands of years.

Supernovae in different cultures

We would expect that ancient astronomers around the world would take special notice of new stars appearing in the night sky. Such observations have come down to us from China, Japan, Korea, the Medieval Islamicate World, and Europe.[1] Written records of astronomical observations have also survived from ancient Greece and Babylon, but the astronomers of these cultures were mainly interested in cyclical phenomena, such as the phases of the Moon, the motions of the planets, and Lunar and Solar eclipses. Some cultures, such as those of the Mayans, Egyptians, and ancient Indians, were known to have a deep interest in astronomy, but no records of their observations have survived to this day. In other cultures, such as those of the Maori,[2] Aboriginal peoples of Australia,[3] or North American native peoples,[4] records of astronomical events may have survived through oral storytelling or material culture (such as rock paintings), but the evidence for observations of supernovae is inconclusive.

The new stars observed by ancient astronomers could have been supernovae, but they could also have been novae, comets, meteors, or even the planets. To decide whether a "new star" mentioned in a historical account refers to a supernova, historians of astronomy look for certain criteria. The new star should have a well-described, fixed position (to rule out planets, comets, and meteors); have no tail (to rule out comets); be exceptionally bright but fade over a long period of time (ideally at least three months); should be close to the plane of the Milky Way (where most of the Galaxy's stars are located); should be associated with a supernova remnant of the same approximate age (these will be discussed at length in chapter 4); and should have several corroborative historical reports. According to these criteria, only five historical events can be counted as real supernovae—the events of 1006, 1054, 1181, 1572, and 1604 CE. Three other candidates—185, 386, and 393 CE—meet some but not all of these criteria.[5]

Richard Stephenson and David Green go into great detail about the historical records and supernova remnants of each of these events in their *Historical Supernovae and Their Remnants*. Here I will concentrate on four examples: the candidate supernova of 185 and the confirmed supernovae of 1006, 1572, and 1604. These examples will show how, over the course of fifteen centuries, the recording of "new stars" spread across the world, from China, Japan, and Korea to the Islamicate World and Renaissance Europe. I will also show how the reason for observing these new stars transformed as well, from astrology to astronomy.

The supernova candidate of 185: the long history of Chinese astronomy

The oldest of the historical supernova candidates, SN 185, is reported in only one trustworthy source. Chapters 20–22 of Sima Biao's (司馬彪) *History of the Later Han Dynasty* (後漢書 *Hou Hanshu*), describe the appearance of a "guest star" (客星 *kexing*):

> Zhongping reign period, second year, tenth lunar month, day 60 [7 December 185 CE]. A guest star appeared in the Southern Gate.[6] It was as large as half a mat, multicolored, and scintillating. It gradually became smaller and in the sixth lunar month of the year after next it vanished. The standard interpretations say this means insurrection. In the sixth year [189 or 190 CE], Yuan Shao, the governor of the Metropolitan Region, punished and eliminated the officials of middle rank. Wu Kuang attacked and killed He Miao, the General of Chariots and Cavalry, and several thousand people were killed.[7]

Most of the Chinese astronomical records follow the above format. Court astronomers observed the night sky regularly and wrote down their observations in log books, which were then summarized in official histories of each dynasty. Sima Biao, for example, compiled his history during the third century CE, the first decades of the Jin dynasty that superseded the Han. Though the official histories were written decades, sometimes centuries, after the fact, they were based on detailed records written at the time of the chronicled events. Each record states when the astronomical event took place, in which Chinese constellation it appeared, and how long it was visible. Finally, the record ends with an astrological interpretation, connecting between the appearance of the guest star and historical events.

This supernova candidate may be the only one to be recorded by the Romans.[8] The reign of Commodus, emperor from 180 CE until his assassination in 192 CE, was plagued by a seemingly never-ending series of calamities: military insurrections in Britain and Gaul, a plague followed by famine, and a fire that destroyed a large swath of Rome, including the sacred temples of Pax and Vesta. The Roman historian Herodianus, who was alive during that time, wrote of this period:

> In that time of crisis a number of divine portents occurred. Stars remained visible during the day; other stars, extending to an enormous length, seemed to be hanging in the middle of the sky. Abnormal animals were born, strange in shape and deformed of limb.[9]

A second excerpt can be found in the chapter dealing with Commodus in the fourth-century *Historia Augusta*:

> The omens that occurred in his reign, both those which concerned the state and those which affected Commodus personally, were as follows. A comet appeared. Footprints of the gods were seen departing the Forum. Before the war of the deserters the heavens were ablaze.[10]

Unfortunately, unlike the Chinese historical records, the Roman records fail to provide the exact date and location of the event and may easily be literary flourishes masquerading as astronomical observations.

The supernova of 1006: the supernova seen around the world

The first confirmed historical supernova, SN 1006, was also the brightest. By the eleventh century, Japanese and Korean astronomers had joined their Chinese colleagues in methodically recording observations of the night sky. Heavily influenced by Chinese culture, the Japanese and Korean records use the same terminology and follow a similar format as the Chinese records.

Visible for three years, the new star was bright enough to be seen in broad daylight. It was described in Chinese records as "huge . . . like a golden disk," "its appearance was like the half Moon and it had pointed rays," and "it was so brilliant that one could really see things clearly (by its light)." A Japanese record compared the brightness of the guest star to that of the planet Mars. The only Korean astronomical record from 1006 CE states: "King Mokchong, ninth year. A broom star was seen." A broom star would usually refer to a comet, but no comets were recorded by Chinese astronomers in 1006 CE, so it is possible that this report actually refers to the supernova.[11]

The new star of 1006 was also noticed by astronomers throughout the Islamicate World, in Egypt, Iraq, northwest Africa, and Yemen. Ali ibn Ridwan (علي إبن رضوان), an Egyptian physician with an interest in astrology, witnessed the new star when he was a young boy. Years later, in his *Commentary on the Tetrabiblos of Ptolemy*, he described the new star as follows:

I shall now describe a spectacle (*athar*) which I saw at the beginning of my studies. This spectacle (*athar*) appeared in the zodiacal sign Scorpio, in opposition to the Sun. The Sun on that day was 15 degrees in Taurus and the spectacle (*nayzak*) in the 15th degree of Scorpio. This spectacle (*nayzak*) was a large circular body, 2.5 to 3 times as large as Venus. The sky was shining because of its light. The intensity of its light was a little more than a quarter of that of moonlight.[12]

Ali ibn Ridwan then goes on to describe the positions of the Sun, Moon, and planets that night, from which it appears that the night in question was that of 30 April 1006 CE, a day before the first Chinese and Japanese detections. It is noteworthy that he describes the new star as either an *athar* (أثر) or a *nayzak* (نيزك)—translated by Bernard Goldstein as "spectacle"—instead of the more common words for "star" in Arabic: *nagem* (نجم) or *kaukab* (كوكب). Other Arabic records describe it as a "great star" (كوكب كبير) of uncharacteristic brilliance (e.g., "its rays on the Earth were like the rays of the Moon").[13]

Two European chronicles, from the monasteries at St. Gallen in Switzerland and at Benevento in Italy, also record the appearance of a new star. The monks at the Benedictine abbey of St. Gallen wrote:

> 1006. A new star of unusual size appeared; it was glittering in appearance and dazzling the eyes, causing alarm. In a wonderful manner it was sometimes contracted, sometimes spread out, and moreover sometimes extinguished. It was seen, nevertheless, for three months in the inmost limits of the south, beyond all the constellations which are seen in the sky.[14]

While the only entry from 1006 CE in the *Annales Beneventani* notes:

> 1006. In the 25th year of Lord Pandolfo and 19th year of his son, Lord Landolfo, a very bright star gleamed forth, and there was a great drought for three months.[15]

Both of these records, as well as the full record written by Ali ibn Ridwan, interpret the appearance of the new star as an ill omen. Some of the Chinese records, on the other hand, refer to the new star as a *Zhoubo* (周伯), which was generally regarded as an auspicious "virtuous star" (德星 *dexing*), as opposed to a "baleful star" (妖星 *yaoxing*).

The large number of records, their spread across East Asia, the Islamicate World, and Europe, and the thorough descriptions of the position of the star allowed modern astronomers to localize it on the sky and connect it to supernova remnant G327.6+14.6.[16]

The supernovae of 1572 and 1604: from astrology to astronomy

The last two examples I will touch upon—the supernovae of 1572 and 1604—mark the transition of supernova studies from astrological to astronomical. As before, the new star of 1572 was recorded by Chinese and Korean observers. While there are no known Japanese or Arabic records, there is a Hebrew record of this new star written by a rabbi from Prague by the name of David Gans (דוד גאנז). Born in 1541, he would have been 31 or 32 when the new star appeared in the night sky. In his chronicle of Jewish and world history, *Zemah David* (צמח דוד), Gans described the new star thus:

> The large and terrible and well-known star with the large tail was seen anew in the year 332, 1572 by the Christian count, and that new star stood for a period of 14 months, longer than any previously known new star mentioned in all the books of recollections. And all the seers in Germany, Italy, France, and Spain aggrandized their prophecies and explanations of what the said star foretold, and wrote many books about it, and prophesied many evils and ruins; God in his mercy frustrates the signs of liars.[17]

It is possible that Gans conflated the new star of 1572, which he notes was visible for an inordinately long period of time, with the Great Comet of 1577, which was visible over Prague for more than two months.[18]

Far more interesting is Gans's description of European observers' attitudes toward the star,[19] and his own criticism—couched in the language of the Bible—of their rush to prophesy. Gans himself often connects between "heavenly signs" (אותות השמים) and earthbound catastrophes; the description of the new star is followed by a tally of three earthquakes that rattled Constantinople, Augsburg, and Munich in 1572, as well as the death of King Sigismund II August of Poland. He thus engages in the same type of astrology he seemingly criticizes. The difference is that his are *postdictions* as opposed to *predictions*, in line with the common Jewish view that "From the day the Temple was destroyed, prophecy was taken away from the prophets and given to fools and children."[20]

David Gans's interest in the new star was shared by several European astronomers, including most famously the Danish astronomer Tycho Brahe, after whom the supernova of 1572 is commonly named. For more than 15 months, Tycho made detailed observations of the new star's position, brightness, and color. His measurements were accurate enough that 370 years later, his description of the new star's evolving brightness was translated into modern measurements and used to confirm it as a supernova, classify its type, and compare it to modern supernovae of the same type.[21]

Far more important were measurements of the new star's position and its distance from neighboring stars (figure 1), taken by Tycho Brahe and other European astronomers, including Thomas Digges (Cambridge), Jerome Muñoz (Valencia), Cornelius Gemma (Louvain), and

Thaddeus Hagecius (Prague). Taking into account the accuracy of sixteenth-century astronomical techniques and devices, these measurements allowed modern astronomers to confidently associate the supernova remnant G120.1+1.4, commonly referred to as "Tycho," with SN 1572.

These measurements also led Tycho Brahe to conclude that the new star had a fixed position relative to nearby stars (unlike the planets, for example, which visibly move across the sky relative to the stars). Tycho described this conclusion in his preliminary report, *On the New Star* (*De nova stella*), published in 1573:

> That it is not in the orbit of Saturn, however, or in that of Jupiter, or in that of Mars, or in that of any one of the other planets, is clear from this fact: after the lapse of six months it had not advanced by its own motion a single minute from that place in which I first saw it. . . . Therefore, this new star is neither in the region of the Element, below the Moon, nor among the orbits of the seven wandering stars, but it is in the eighth sphere, among the other fixed stars. . . . I conclude, therefore, that this star is not some kind of comet or a fiery meteor, whether these be generated beneath the Moon or above the Moon, but that it is a star shining in the firmament itself— one that has never previously been seen before our time, in any age since the beginning of the world.[22]

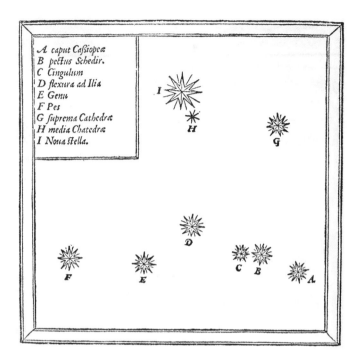

Figure 1 A star chart of the Cassiopeia constellation, drawn by Tycho Brahe. The supernova of 1572 is marked by the letter I and denoted as a new star ("Nova stella"). Source: T. Brahe, *De nova stella* (Copenhagen: impressit Laurentius, 1573).

The "Element" and the various "spheres" refer to Aristotle's cosmology, according to which the heavens are divided into concentric spheres, each containing a different set of celestial objects. One of Aristotle's assertions was that any change in the heavens could only occur in the sub-Lunar sphere. Tycho Brahe's conclusion that the new star had to be located among the fixed stars meant that Aristotle's cosmology, the reigning paradigm in Europe and the Islamicate World for nearly two thousand years, was wrong. Coming on the heels of Nicolaus Copernicus's 1543 *On the Revolutions of the Heavenly Spheres* (*De revolutionibus orbium coelestium*), in which he made the argument that the Earth revolved around the Sun, the observations and analysis of SN 1572 played a significant role in the scientific revolutions that led to the birth of modern astronomy.

The perception of the cosmos as an immutable celestial sphere came under further attack with the appearance of the supernova of 1604. Extensively observed by Chinese, Korean, and European astronomers, it is popularly referred to as "Kepler's supernova" after Johannes Kepler, who made detailed observations of the new star (figure 2). Kepler, who had been Tycho Brahe's assistant for a short time, came to the same conclusion as his late supervisor: that the new star's position did not change and so had to lie among the fixed stars.

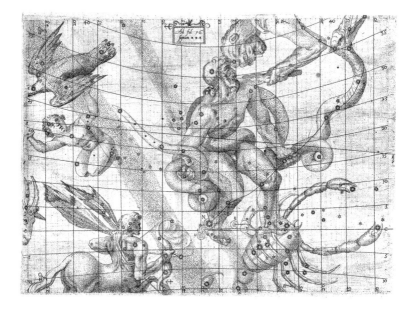

Figure 2 The supernova of 1604, marked by the letter N, appeared in the heel of the constellation Ophiuchus, the "serpent bearer." Source: J. Kepler, *De stella nova in pede serpentarii* (Prague: Typis Pauli Sessii, 1606).

The supernova of 1604 represents another watershed moment in the history of astronomy. Galileo Galilei was among the European astronomers who observed the supernova, and he arrived at the same conclusion as Kepler. These observations, as well as all the other observations described in this chapter, were conducted with the naked eye. In 1609, five years after the appearance of SN 1604,

Galileo would construct his first telescope, based on a Dutch design, and train it on the heavens. The magnification afforded by his telescope led Galileo to describe mountains on the Moon, the phases of Venus, rings around Saturn, and four moons orbiting Jupiter.[23] The modern era of astronomy had begun.

SUPERNOVAE IN THE MODERN ERA

The conversation around "new stars" has continued unabated. Tycho Brahe's "stella nova" was eventually shortened to "nova" (plural: novae) and used to describe any starlike object that appeared suddenly in the night sky and eventually faded away and disappeared. Novae were routinely discovered and reported by astronomers around the world right up to the twentieth century. At the same time, telescopes became steadily more powerful, and the way astronomers used them changed as well.

The age of the telescope

Galileo's telescope was a "refracting" telescope (or "refractor"); it was composed of two lenses that refracted incoming light (i.e., changed the angle at which it traveled) and focused it onto his eye. A telescope is characterized by the diameter of its aperture, the opening through which light enters, which in turn is set by the diameter of the lenses. The bigger a telescope's aperture, the more light can enter, which makes it possible to see fainter objects. This is why astronomers strive to build ever-bigger telescopes. However, in refracting telescopes, using larger lenses also requires using longer tubes. This becomes structurally prohibitive as the lenses grow larger and more massive. The biggest refracting telescope has a diameter of roughly 1 meter. Built in 1897, it is still in operation at Yerkes Observatory.

The development of refracting telescopes was mostly abandoned during the twentieth century in favor of a different kind of telescope. "Reflecting" telescopes (or "reflectors") use mirrors instead of lenses. Unlike refracting telescopes, the length of the telescope does not depend on the size of the mirror as much,[1] so while refracting telescopes are longer than they are wide, reflecting telescopes are the opposite. The largest reflecting telescopes currently in operation have mirrors with diameters of 10–11 meters but a stocky build.

The way we use telescopes has changed as well. On dark nights, Galileo would peer through his telescope and sketch what he saw; his observations were works of art. If you had a telescope growing up, this is probably how you used it as well. Squinting through the eyepiece, you could make out features on the Moon, see the planets and some of their moons, and on very dark nights maybe see some star clusters or the nearest galaxies.

These kinds of observations are limited by the physiology of our eyes, dooming naked-eye observations to the brightest objects in the sky. For one thing, our eyes' pupils, much smaller than telescope apertures, let in only a relatively small amount of light. For another, while modern astronomy cameras can collect light for hours at a time, the nerve cells in our retinas limit the light collection time of our eyes to just a fraction of a second.

In the nineteenth century, the astronomer's eye was replaced by a camera. The light entering the telescope was focused onto a strip of film or a glass plate coated with an emulsion of silver halide crystals. When exposed to light, the crystals underwent a chemical reaction that varied in intensity with the amount of light impinging on the emulsion. The exposed plates or films could then be chemically developed to create black-and-white images. Astronomers could now take longer exposures and begin to make out objects too faint for our eyes to see, such as "new stars" in other galaxies. These observations set the path for some "novae" to become "supernovae."

From novae to supernovae

In the short period of calm between the two world wars, astronomers began to classify novae into lower-, middle-, and upper-class novae based on their brightness. Then, in a long-forgotten paper from 1932, Swedish astronomer Knut Lundmark used a new appellation for the "upper-class" novae: *super-novae*.[2] As far as I can tell, this marks the first appearance of the word (at least in English).[3]

At the time, the distinction between the "middle-class" and "upper-class" novae rested on four objects: the "new stars" observed in 1572 and 1604 (chapter 2), as well as similar phenomena that had appeared in the nearby galaxies Andromeda in 1885 (S Andromeda or SN 1885A) and M101 in 1909 (known as SS Uma or SN 1909A).[4] It is a common cliché in modern astronomy that "one abnormal object is an outlier; two, a new class," but it is still breathtaking when a whole new type of astronomical phenomenon, and one that will become an entire field of study, is claimed on the basis of no more than four objects.

In 1934, super-novae were formally split off from their fainter nova cousins. That year, Walter Baade and Fritz Zwicky,[5] astronomers at Mount Wilson Observatory and Caltech, respectively, published two ground-breaking papers. In the first, "On Super-Novae," they used simple energy calculations to conclude that a star experiencing a super-nova would lose a large fraction of its mass.[6] They even speculated that what would be left behind—the remnant—could be a neutron star. This was an inspired claim, as the neutron had only been discovered two years earlier.[7] The idea that an entire star could be composed of this new particle must have sounded preposterous (like many of Zwicky's claims).

Baade and Zwicky did not stop at two revolutionary ideas. In their second paper, "Cosmic Rays from Super-Novae," they suggested that these energetic explosions could be the cause of the enigmatic "cosmic rays," a source of radiation in the Earth's atmosphere that was revealed to be cosmic in origin by Victor Hess in his 1911 and 1912 balloon experiments (chapter 5).[8]

Although it took decades for Baade and Zwicky's speculations about neutron stars and cosmic rays to be accepted by the astronomical community, their papers solidified the distinction between common novae and super-novae. By 1938, the hyphen in "super-novae" had disappeared and "supernovae" were no longer grouped together with novae. However, while Baade and Zwicky's papers already postulated that a supernova would cause the complete destruction of the star, it would be forty more years before astronomers understood common novae (now simply referred to as **novae**) to be the result of thermonuclear explosions on the surfaces of certain types of stars—explosions strong enough to be seen across the expanse between galaxies but not strong enough to fully destroy the star.[9]

The dawn of supernova surveys

Following the publication of these seminal papers, Fritz Zwicky set out to discover supernovae in a systematic way.

His first supernova survey, with a 3-inch-diameter telescope on the roof of the Astrophysics building at Caltech, was a failure due to the poor observing conditions. In 1935, he convinced the Council for the Construction of the Palomar Observatory to build him an 18-inch (roughly half-meter-diameter) telescope for the sole purpose of searching for supernovae. The telescope was inaugurated ("saw first light" in astrophysical jargon) on the night of 5 September 1936. By the end of this survey in December 1940, Zwicky and his collaborator, Josef J. Johnson, had discovered nineteen new supernovae.

Zwicky and Johnson quickly searched each new image the day after it was taken. Each new image had to be painstakingly compared by eye to previous images of the same patch of sky in order to look for new points of light. At the same time, this process had to be done quickly to ensure that any new supernovae could immediately be observed by the more powerful 200-inch (5-meter) telescope.

Zwicky paused his survey during 1941–1954 due to World War II and the production of a catalog of 40,000 bright galaxies and 10,000 galaxy clusters. He restarted the survey in 1954, and in 1957 several astronomers at observatories around the world started searching for supernovae as well. This international push led Zwicky to propose in 1961 that the International Astronomical Union create a Committee for Research on Supernovae within Commission 28 (in charge of "extragalactic nebulae," the previous term for galaxies). The first order of business of the new committee was to organize an international survey that included telescopes in Argentina, France, Italy, Mexico, the USSR, and the USA.

The transition to an organized, international survey quickly bore fruit. Zwicky noted that 54 supernovae had been discovered during the 70 years between 1885 and 1956. Between 1956 and 1962, in the span of only six years, the supernova surveyors spread around the world had doubled this tally with the discovery of 60 new supernovae.[10]

Today: digital cameras and international collaborations

In these pioneering surveys, Zwicky and his collaborators laid the foundations of the modern field of supernova research. Except for matters of scale and technology, today's supernova surveys are almost indistinguishable from Zwicky's surveys. A modern survey is usually undertaken by a group ("collaboration") of astronomers with access to time on one or more telescopes. Some of the telescopes are used to discover supernova candidates, while others are used for follow-up observations of the candidates to chart how they brighten and fade over time. A recent example is the Zwicky Transient Facility, a supernova survey run out of Palomar Observatory (figure 3). The observatory's

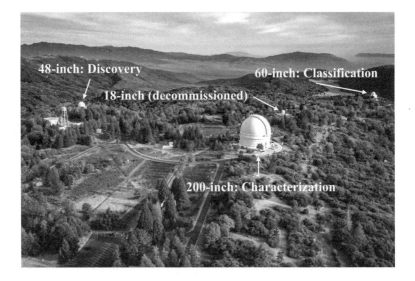

Figure 3 Aerial view of Palomar Observatory. The telescopes used by the Zwicky Transient Facility survey are highlighted with arrows. The 48-inch telescope is used to discover candidates. Classification and further characterization of the supernovae are then conducted with the 60- and 200-inch telescopes. The dome that housed the now-decommissioned 18-inch telescope used by Zwicky is also highlighted. Credit: Caltech/ Palomar Observatory.

48-inch telescope is used to discover supernova candidates, which are then classified with the 60-inch telescope. The 200-inch telescope (which Zwicky used to classify the candidates he discovered with his 18-inch telescope), along with telescopes at other observatories around the world, then conduct follow-up observations to further characterize the new supernovae.

Figure 4 The Hubble Space Telescope, in orbit around Earth. Credit: NASA/ESA.

As in all of astrophysics, over the last fifty years the telescopes used for supernova surveys have significantly increased in size, from Zwicky's 18-inch telescope to today's 10-meter telescopes, a nearly 500-fold increase in light collection area.[11] Although most of our telescopes are still built on remote mountains, we also routinely use space telescopes in orbit around the Earth, such as the famous 2.4-meter Hubble Space Telescope (figure 4). Perhaps the most important change, though, has been the switch from photographic plates to charge-coupled devices (commonly referred to by their initials, CCDs).

In order to discover a supernova, Zwicky and his assistants would first develop the film exposed by the 18-inch telescope, then lay the new film over an older one, usually taken a month before. They would carefully align the two pieces of film and then slightly shift one of them, so that each star in the new film appeared next to the same star in the older film. A supernova candidate would appear as a lone dot on the new film.[12] In later years, Zwicky also added the 48-inch telescope on Palomar Mountain to his survey. Unlike the 18-inch, the 48-inch used glass plates. These, too, were developed on site and then slotted into a blink stereo comparator microscope, in which an optical assembly was used to periodically alternate between the images cast by two different plates, so that a point of light that appeared on one plate but not the other would appear to blink in and out.[13]

These methods were slow, laborious, and highly dependent on the skill of the astronomer searching the images. It is easy to imagine Zwicky and his assistants, after a night at the telescope, hunched over a new film they had just taken and an older film of the same galaxy, missing the appearance of one small point of light on either of the films. In a 1938 paper, Zwicky referred to this problem as "the personal equation."[14]

The CCD, invented in 1969 by Willard S. Boyle and George E. Smith at AT&T Bell Labs,[15] now lies at the heart of most of the cameras used for astronomy. A CCD is an integrated circuit chip composed of a grid of light-sensitive sections called "pixels" (short for "picture elements"). When light is focused onto a CCD, each pixel begins to accumulate an electric charge due to the photoelectric effect. Once the CCD is no longer exposed to the light source, the electric charge in each pixel is transferred sideways along the grid until it reaches the edge of the chip. There, it is converted into a digital signal and read out.

In the final digital image, the electric charge from each pixel is represented by a shade of gray. Areas with low illumination will be composed of pixels with darker shades of gray; areas with greater illumination with lighter shades. In this respect, a CCD image is similar to a photograph captured on film.

In old analog images, an object's brightness was estimated by comparing it to other objects nearby and to well-known objects that served as benchmarks. To the naked eye, and on photographic films or plates, brighter objects look bigger. This led the ancient Greeks to devise the "magnitude" system,[16] a variation of which is still in use today. According to this system, the star Vega has a magnitude of zero. Brighter stars, like our Sun, will have negative values, while fainter stars will have positive values. The magnitude system is logarithmic, so that a −1 magnitude star is roughly two and a half times brighter than a 0 magnitude star, and a –2 magnitude star is about six times brighter.

For two thousand years, astronomers would look up at the sky, or down at a glass plate, and would say: "this star looks brighter than that one but fainter than that other one, so its brightness is roughly 5 magnitudes." Though it served the field well, this system was subjective and not very precise.

The CCD made it easy to manipulate images and make high-precision measurements of brightness. Each pixel in a CCD image records the amount of electric charge liberated by the light that hit that particular part of the chip. The final CCD image is a grid of those numbers. A star will occupy a certain number of pixels (it is still true that a brighter star will appear bigger—and take up more pixels—than a fainter one), and the values of those pixels can be summed up and converted into a precise measurement of the star's brightness.

Moreover, different images of the same area on the sky can be combined using a computer (which aligns the images and then sums them up). This results in deeper images, which reveal fainter objects that would have been invisible in each of the separate exposures.

For historical reasons, we still use the magnitude system (though some fields of astronomy have moved away from it), but now we can say: "this star has 500 ± 90 counts, which means it has a brightness of 5.2 ± 0.2 magnitudes."

The second number, after the plus-minus sign, is the uncertainty of the measurement, which depends on such factors as the duration of the exposure, the engineering specifications of the CCD chip, and the brightness of the star. Crucially, that number no longer depends on the skill or experience of the astronomer taking the measurement. The introduction of the CCD transformed astronomy into a modern, computerized science and took some of the bite out of Zwicky's "personal equation."

Another important difference between Zwicky's and modern surveys concerns the types of targets aimed at by the surveyors. In today's jargon, Zwicky's survey would be labeled "targeted," since he chose to visit and revisit a predetermined list of about 3,000 target galaxies every month. By observing these galaxies once a month for years on end, Zwicky could monitor them closely and discover new supernovae that exploded in them.

Targeted surveys have lost favor in recent years, since they result in biased supernova samples, in which the only types of supernovae that can be discovered are those that are common in the types of galaxies that were preselected by the observer. Because supernovae are explosions of stars, to maximize the output of a targeted survey it makes the most sense to observe the largest galaxies; more stars will net more supernovae. However, we now know that some types of supernovae only occur in small, so-called "dwarf" galaxies. Though they host fewer stars, those stars have a different chemical composition than the older stars commonly seen in large galaxies. Thus, if you only target large galaxies, you might find more supernovae but you will miss the ones that explode in other types of galaxies. The last major "targeted" survey of this type, the Lick Observatory Supernova Search, began in March 1998. In its first decade of operation, it repeatedly observed nearly 15,000 bright, massive galaxies in our immediate galactic neighborhood, and discovered more than a thousand supernovae.[17]

Most present-day supernova surveys are "untargeted." Instead of periodically visiting a set list of galaxies, an untargeted survey will periodically take an image of a certain patch of sky and each of the galaxies in that patch—large and small galaxies alike—will be searched for supernovae.

The transition from targeted to untargeted surveys is also a result of the introduction of CCDs. Depending on the field of view of the camera used for the survey, a single image may include between dozens and thousands of galaxies. Instead of searching each new image by eye and trying to locate a new point of light among the thousands of objects in the image, an older image is subtracted from the new one and only the "difference" image is searched. This is made possible by the digital nature of images produced by CCD cameras. Whereas glass plates or strips of film were physical objects that had to be placed next to one another, a CCD image is a grid of numbers that can be manipulated by computer algorithms.

The process of subtracting one image from another is straightforward; the number in one pixel is subtracted from the number in the same pixel in the other image. The resulting difference image is a representation of the new value in each pixel. Most astronomical objects, such as stars and galaxies, change on timescales of millions or billions of years. In two images taken a few days, weeks, or months apart those objects will register very little change. So, while they will be visible in the new ("target") and old ("template") images, they will disappear in the difference image.

On the other hand, supernovae and other variable objects, which do change on short timescales, will appear in the difference image and be easy to pick out. Other objects, such as comets or asteroids, which change location from one image to the next, will also appear in the difference image, as will cosmic rays. Thus, even though difference imaging makes it easier to search for supernovae, there is still an art to it. The digital nature of CCD images makes it possible for computers to search difference images for supernovae. This alleviates some of the burden of the "personal equation" but does not vanquish it completely; the algorithms we use to produce and search difference images are still written by humans and so are far from perfect.

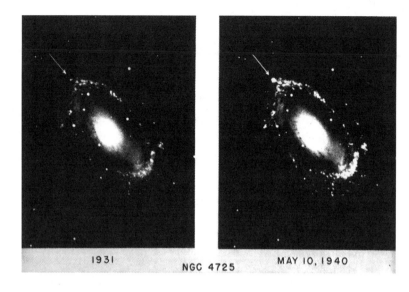

Figure 5 SN 1940B, discovered by Josef J. Johnson during Fritz Zwicky's inaugural supernova survey of 1936–1940. The left- and right-hand panels show images of the host galaxy, NGC 4725, taken before and after the star exploded. In each panel, the location of the supernova is indicated by a white arrow. Source: F. Zwicky, "Supernovae," in *Stellar Structure*, ed. L. Aller and D. McLaughlin (Chicago: Chicago University Press, 1965).

Figures 5 and 6 show the progression from Zwicky's inaugural survey and his use of film to today's surveys and CCD-produced difference images. In figure 5, one of the nearby galaxies monitored by Zwicky (NGC 4725) with the 18-inch telescope is shown in two films taken in 1931 and 1940, respectively.[18] By comparing them, one notices the bright new star in the image from 1940, highlighted by the white arrow.

Figure 6 (recreated in color in plate 2) shows two supernovae discovered in images taken with the Hubble Space Telescope in 2019. A close inspection of these CCD images will reveal their pixelated nature. The rightmost panels are difference images created by subtracting "template" images taken in 2016 from "target" images taken in 2019. The difference images are mostly dark because all of the stars and galaxies, which have not changed much over this span of time, are subtracted out. The supernova BUF19Awd, whose light reached us in 2019, appears as a bright point of light in the difference image, while BUF-19Sop, which disappeared between 2016 and 2019, appears as a dark "hole" in its difference image.[19]

While Zwicky targeted nearby galaxies, the two galaxies shown in figure 6 are just two galaxies out of hundreds in the fields imaged by the Hubble Space Telescope (plate 2). These galaxies are also farther away than the ones viewed by Zwicky; in his images, they would have appeared as nothing more than small, faint blobs. The supernovae would not have been detected at all.

For more than 1,800 years, astronomers around the world have observed the appearance of new stars in the night sky. At first, these luminous visitors were used for divination. Then, as modern science emerged in the sixteenth century and the telescope in the seventeenth, the interest in new stars gradually shifted from astrology to astronomy. The twentieth century brought several

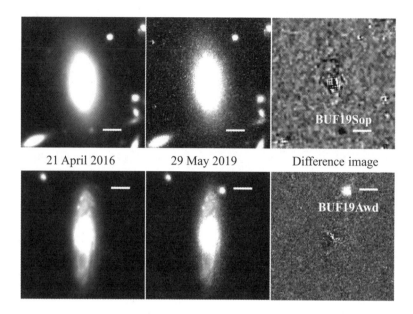

| 21 April 2016 | 29 May 2019 | Difference image |

Figure 6 Two supernovae discovered in the Hubble Space Telescope Beyond Ultra-deep Frontier Fields and Legacy Observations (BUFFALO) survey. In each row, the image taken in 2016 (left) is digitally subtracted from an image taken in 2019 (center) to produce a difference image (right) in which the galaxies and stars have disappeared and only the supernova candidate remains. BUF19Sop and BUF19Awd appear as a dark "hole" and a bright spot in their respective difference images. See plate 2 for an extended color version. Credit: O. Graur (U. Portsmouth), A. M. Koekemoer (STScI), C. L. Steinhardt (U. Copenhagen), and M. Jauzac (Durham U.).

Today, hundreds of astronomers routinely discover thousands of supernovae each year.

revolutions to this field: a conceptual shift that divided the family of new stars into "novae" and "supernovae," followed by leaps in camera technology, which allowed astronomers to systematically search for supernovae in all types of galaxies instead of waiting for them to appear in our own every few centuries. Today, hundreds of astronomers routinely discover thousands of supernovae each year in a panoply of surveys. The following chapters describe what these observations have taught us about the physics of supernovae and how we have come to use them as tools in other experiments.

ENLIGHTENED BY LIGHT

The emergence of powerful telescopes and cameras made it possible to methodically discover and study supernovae. The relatively new field of supernova studies has been a busy one; over the last few decades, our knowledge and understanding of these explosions has expanded significantly. However, before we delve into *what* we know about supernovae, we need to understand *how* we come to know it.

Light

Most of our knowledge of the Universe derives from the light we collect with our telescopes. In physics, light is treated as both a wave and a particle. As a particle (called a "photon"), we usually describe light in terms of

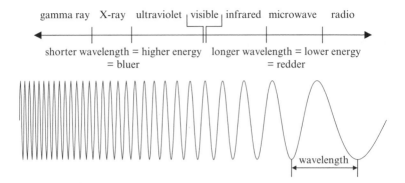

Figure 7 The electromagnetic spectrum. Different types of light have longer or shorter wavelengths. Our eyes only perceive light in the "visible" band, a thin sliver of the electromagnetic spectrum.

energy: "high-energy" or "low-energy" photons emitted by a source. As a wave, light is a disturbance of electric and magnetic fields (which is why it is also referred to as an "electromagnetic" wave). The most important characteristic of any wave is its **wavelength**, which describes the width of the wave between two consecutive peaks (figure 7). Wavelength can also be thought of in terms of frequency or energy: waves with long wavelengths will have lower frequencies and lower energies (in other words, a high-energy photon is analogous to a light wave with a short wavelength).[1] To our eyes, light with a longer wavelength will also appear redder than light with a shorter wavelength, which will appear bluer.

Figure 7 also shows the electromagnetic spectrum, which describes our division of light, according to wavelength, into different bands; radio and infrared light have longer wavelengths than ultraviolet, X-rays, or gamma rays. Sadly, the light our eyes can process, called "visible" or "optical" light, is just a tiny sliver of this spectrum. Zwicky's surveys, described in chapter 2, were limited to this narrow range. But modern telescopes have taken the blinkers off our eyes. Today, specially designed telescopes and cameras make it possible to view the Universe—and the supernovae in it—in all wavelengths, from radio to gamma rays.

Supernova classification

"One of these things is not like the others . . ." Like toddlers watching *Sesame Street*, one of the first things scientists do when studying a group of objects or phenomena is to sort the objects they study, be they butterflies or supernovae, into different categories.

If you set out to collect butterflies, you will quickly discover that they come in a wide range of sizes, shapes, and colors. Look closer and you will see that some of the specimens in your collection have a variety of antennae; some are threadlike and some are comblike, some are thick and some are slender, some are long and some are short.

A further investigation will reveal that all the specimens with clubbed antennae tend to fold their wings vertically up over their backs, while all the specimens with feathery or saw-edged antennae tend to hold their wings in a tent-like fashion that hides the abdomen. We call the former butterflies and the latter moths.

Once differentiated, it is easier to discover other differences between the two groups and perhaps divide them into smaller and smaller subgroups. This classification process helps scientists make sense of their observations and leads to more specific questions suited to each group, such as "why do moths need different antennae than butterflies?" The same is true for supernovae.

In 1941, Rudolph Minkowski used a sample of 14 objects to divide the nascent supernova class into objects of Type I and Type II.[2] As astronomers discovered more and more supernovae, further divisions were suggested, argued over, and finally adopted or discarded.[3]

The current classification scheme is shown in figure 8.[4] It is mostly based on two sources of information: **light curves** and **spectra**. A supernova's light curve (figure 9) is composed of measurements of its brightness taken over time with various filters (the physics powering the light curve is described in detail near the end of this chapter). A spectrum, obtained with an instrument called a "spectrograph," reveals how much light is emitted by the object in each wavelength *individually*. Rainbows are naturally

Type II: Hydrogen-rich, core-collapse explosions

IIP	100-day-long plateau ("P") after light curve peak.
IIL	Light curve fades in a linear ("L") manner after peak.
IIn	Narrow ("n") hydrogen emission features.
II-87A	Long rise to peak (>80 days).
IIb	Hydrogen emission features disappear over time.

Type Ib/c: Hydrogen-poor, "stripped-envelope" core-collapse events

Ib	Broad helium spectroscopic features.
Ibn	Narrow ("n") helium emission features.
Ic	No hydrogen or helium features in spectra.
Ic-BL	Very broad spectroscopic features (BL: broad lines). Some Ic-BL associated with gamma-ray bursts.

Type Ia: Hydrogen-free, thermonuclear white-dwarf explosions

Ia-Normal	Prominent silicon features, two peaks in infrared.
Ia-91T	More luminous than Normal but similar spectra.
Ian	More luminous than Normal, narrow hydrogen features.
Ia-91bg	Dimmer than Normal, no second infrared peak.
Iax	Dimmer than Ia-91bg, narrower spectroscopic features.
Ca-rich	Dim explosions with strong calcium features.

Exotic: Recently discovered, poorly understood explosions

Superluminous	Most luminous. Types I and II: without / with hydrogen.
Rapidly evolving	Rapid (<10 days) rise to peak and rapid decline.

Figure 8 The supernova classification scheme. Only the most common supernova types are shown. The exotic supernovae are described in detail in chapter 7.

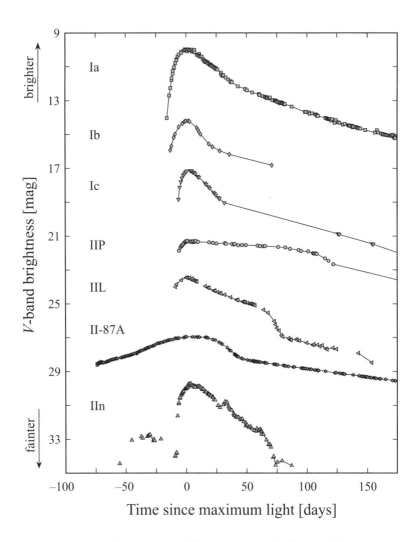

Figure 9 Examples of supernova light curves in the *V* band (comparable to green light). For clarity, all light curves except for that of the Type Ia SN 2011fe have been shifted vertically. Day zero corresponds to the time of peak as measured in the *B* (blue) band. Some Type IIn supernovae, such as SN 2009ip shown here, have pre-supernova outbursts years to days before the main explosion.[5]

occurring spectra (plural of "spectrum") that are created when light (usually, but no limited to, sunlight) is refracted by raindrops.

The spectrum of any astrophysical object is composed of three components. The overall shape of the spectrum—its "continuum"—is due to light emitted by the object across all wavelengths. On top of the continuum, a spectrum can also include emission or absorption features. An emission feature, which looks like either a narrow spike or a broad hill, is created when an atom of a given element (such as hydrogen or silicon) transitions from a high energy state to a lower one by emitting a photon. Conversely, an absorption feature, which looks like an inverted spike or a wide trough, is created when an atom absorbs a photon and removes it from the continuum. Figure 10 shows examples of spectra taken of various supernovae and the telltale signs (or lack thereof) of the elements used to classify their type.

With time, we have attempted to extract every bit of information available in these sources: the shapes and widths of the absorption and emission features—the dips and spikes—tell us about the chemical composition, velocity, and distribution of the material ejected by the supernova, while the shapes of the light curves hold hints to the explosion physics, the sizes of the exploding stars, and much more.

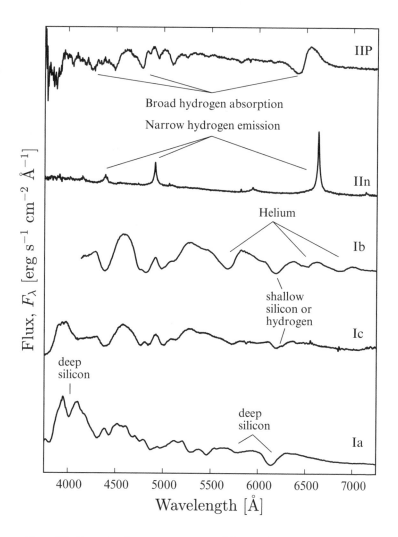

Figure 10 Examples of supernova spectra and the main features used for classification.[6]

The classification scheme shown in figure 8 is still in flux; by the time you read this chapter, new subtypes may have been suggested or older subtypes merged.[7]

Light curve and spectroscopic physics

The light curves of supernovae are powered by several physical processes working in tandem with the expansion of the cloud of debris—called the supernova **ejecta**—that used to be the star. This cloud expands into space at thousands to tens of thousands of kilometers per second—a few percent the speed of light.

In many supernova types, the earliest light received from the explosion is due to "shock breakout."[8] This occurs when the blast wave that blew up the star (chapter 4) reaches the edge of the ejecta and produces a brief, seconds-long flash of X-rays and ultraviolet light. The cooling ejecta then proceed to emit ultraviolet and optical light for about a day. At this point, the rest of the ejecta is still too dense to allow light generated deeper within the star to escape, and the light curve drops significantly. Because shock breakouts are brief and occur shortly after explosion, only a small number have been observed.[9]

As the shock travels through the star, some of its energy goes into propelling the ejecta outward while another fraction goes into heating it. The heated gas then cools down by emitting photons. At first, these photons are trapped inside the dense ejecta, but as the supernova ejecta continue to expand into space, their density decreases, allowing those photons to stream out. The ejecta cool from the outside in, leading to a buildup of light emitted from successive layers. This causes the light curve to rise quickly, as seen in the Type IIP and IIL light curves in figure 9.

In Type IIP supernovae, a coincidence between the rate at which the shocked hydrogen in the ejecta cools and emits photons and the velocity at which the ejecta expand and allow those photons to stream out produces a long-lasting plateau phase (hence the "P" in IIP) dominated by broad hydrogen features. Once the ejecta have cooled, Type II light curves suffer a sharp drop in brightness before transitioning to being powered by radioactive decay of elements created during the explosion. The lack of a plateau in Type IIL supernovae—where "L" stands for the linear drop in their brightness—is thought to be due to a thinner envelope of hydrogen.

In Type I and IIb supernovae, the entire light curve is powered by the radioactive decay of an unstable isotope of nickel that is formed during the explosion. ^{56}Ni (pronounced "nickel fifty-six"), which is composed of 28 protons and 28 neutrons, decays into an unstable form of cobalt, ^{56}Co, which in turn decays into stable iron, ^{56}Fe.[10] Most of the radioactive nickel is synthesized deep inside the star, where conditions are dense enough for its creation. The photons emitted by its decay into ^{56}Co have to make their way through the expanding ejecta, lengthening the rise of the light curve. Once all of the photons can escape the ejecta freely, the shape of the light curve is determined by the dwindling supply of radioactive nickel, and the supernova enters a long decline phase that lasts for years.[11]

In Type IIn supernovae, the post-shock emission is due to interaction between the supernova ejecta and shells of gas in the immediate vicinity of the supernova. These gas shells were blown away by the star in winds or outbursts prior to the explosion (chapter 4).[12] As the ejecta expand, they may encounter successive shells of expelled gas, and the light curve may exhibit successive peaks. The surrounding gas, heated by the interaction with the expanding ejecta, will cool down by emitting light. This manifests in narrow (hence the "n" in IIn) emission peaks in the spectra.

The same processes that shape the light curves of the supernovae also affect the appearance of their spectra. In the first, "photospheric" phase, the slow diffusion of photons through the hot, thick ejecta manifests in a blue, featureless continuum. As the ejecta expand and reveal the inner layers of the blown-up star, the spectra begin to exhibit absorption features that reveal the star's composition. Once the ejecta have expanded sufficiently for light to stream out freely, the supernova enters a **nebular** phase during which the spectra only exhibit emission features.

The peak luminosity of supernovae—how high their light curves rise before beginning to decline—is also governed by the processes described above. In supernovae powered by radioactive decay, the peak luminosity is set by the amount of ^{56}Ni produced during the explosion. In Type Ia supernovae, roughly half the exploding star is converted into ^{56}Ni.[13] In most Type II and Ib/c supernovae, only about 1% of the star's mass is converted into ^{56}Ni, which is why these supernovae are on average less luminous than Type Ia supernovae.[14] The luminosity of Type IIn supernovae depends on the amount of matter encountered by the supernova ejecta and varies widely, with some events as faint as Type IIP supernovae and others brighter than Type Ia supernovae. In all cases, a supernova at peak will be one of the brightest objects in its host galaxy (plate 1). And during its first month, a supernova will emit more light than all of the billions of stars in its host galaxy combined.

Supernova rates

Supernovae are a common phenomenon; every second, about seventy supernovae explode somewhere in the Universe.[15] The rate—i.e., frequency—at which galaxies produce stars has changed throughout cosmic history. As

During its first month,
a supernova will emit
more light than all of
the billions of stars in its
host galaxy combined.

galaxies started forming and growing in size, the star-formation rate rose until it peaked some ten billion years ago. Since then, the rate at which new stars are born—and explode—has been steadily declining.[16]

Because of their higher luminosities, Type Ia supernovae are the most common type of supernova detected in surveys. However, Type II supernovae are actually the most common explosion in nature, accounting for half of all supernovae in the Universe today. A quarter of all supernovae are Type Ia, and Type Ib/c supernovae account for the final quarter.[17]

A common misconception is that a galaxy will experience one supernova per century. In fact, more massive galaxies, which host more stars, will experience higher rates of supernova explosions. For example, in NGC 6946, which is roughly ten times more massive than our Galaxy, ten supernovae have been observed in the last 100 years alone,[18] earning it the nickname "the Fireworks galaxy."

Given the size and mass of the Milky Way Galaxy, we expect it to host 2–3 supernovae per century.[19] Yet four hundred years have elapsed since we last observed a Galactic supernova (SN 1604). We attribute this dearth of local supernovae to two factors. First, the rate quoted above is the *average* rate, which means that some centuries will boast more than three explosions while other centuries will have fewer than two or none at all. Second, Earth is located in an outer arm of the Galaxy, which means that

Supernovae are a common phenomenon; every second, about seventy supernovae explode somewhere in the Universe.

Given the size and mass of the Milky Way Galaxy, we expect it to host 2–3 supernovae per century.

we, its inhabitants, view most of the Galaxy through a thick screen of dust, which absorbs light and either dims or completely obscures our view (for more on dust, see chapter 6).

At least two supernovae exploded in our Galaxy during the intervening four centuries, but no historical records of them have been discovered. We know of them from the existence of two supernova remnants aged ~300 years (Cassiopeia A) and ~140 years (G1.9+0.3).[20] Obscuration by dust could explain why the supernovae that left these remnants were never observed. The number of known supernova remnants in our Galaxy is also lower than expected given the supernova rate and the way supernova remnants shine over time. Discovering more remnants, especially young ones, will reveal other supernovae we might have missed as we bide our time until the next Galactic supernova lights up the night sky.

In the meantime, we can use the various tools at our disposal to poke and prod the supernovae we observe until they give up their secrets. In the next chapter, I describe what we have learned so far about the nature of these brilliant explosions by delving into the fascinating lives and deaths of stars.

THE LIVES AND DEATHS OF STARS

Supernovae are explosions of stars at the end of their lives. So, to understand how supernovae occur—how stars die—we first need to understand how stars are born and develop over time.

A star is born

The Universe was not always full of stars. Even today, it is mostly filled with hydrogen atoms separated by vast empty tracts of space. Interspersed among these hydrogen atoms are a few helium atoms and a minuscule number of atoms of all other elements. In numbers, hydrogen accounts for ~73% of the visible mass in the Universe, helium takes up ~25%, and the remaining <2% comprises all the other elements in the periodic table, from lithium to uranium (see chapter 5 for further details).

Over millions of years, the gravitational attraction between these atoms, though weak, pulls them together to form **molecular clouds**—areas of space that are just slightly denser than the surrounding space, yet dense enough that the disparate hydrogen atoms can clump together to form H_2 molecules (pronounced "H-two," these are molecules made up of two hydrogen atoms). As these clouds of hydrogen slowly contract, they break up into smaller clumps that continue to shrink under the influence of gravity.

When a gas contracts in volume, its pressure and temperature rise in turn. The pressure exerted by the movement of the heated gas molecules (gas pressure) eventually grows strong enough to counteract the gravitational pressure and halt any further collapse of the cloud (figure 11).

Molecular clouds less massive than ~10% the mass of our Sun stop evolving at this point. The resultant stars, called **brown dwarfs**, smolder in infrared wavelengths as they slowly radiate away the heat built up by the cloud's collapse, eventually fading to black.

In more massive clouds, the pressure and temperature of the gas also rise until the emergent gas pressure halts the gravitational collapse. However, in these clouds, the temperature and pressure in the innermost part of the nascent star—its core—rise high enough to also induce the

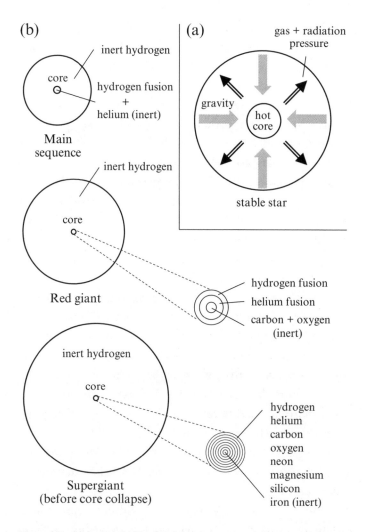

(b)

inert hydrogen

core

hydrogen fusion
+
helium (inert)

Main
sequence

(a)

gas + radiation
pressure

gravity

hot
core

stable star

inert hydrogen

core

Red giant

hydrogen fusion
helium fusion
carbon + oxygen
(inert)

inert hydrogen

core

Supergiant
(before core collapse)

hydrogen
helium
carbon
oxygen
neon
magnesium
silicon
iron (inert)

Figure 11 **(a)** A star is stable when its constituent gas is hot enough to exert an outward gas pressure that counteracts the inward gravitational pressure. **(b)** In each successive phase of a star's life, its core fuses heavier and heavier elements, ending with the creation of iron.

nuclear fusion of hydrogen into helium.[1] Light created by the fusion process steadily makes its way out of the core and through the outer envelopes of inert gas.[2] The hot, dense ball of gas begins to shine and a star is born.

The nuclear reactions in the star's core create a thermostat-like effect that both stabilizes the star and regulates its temperature and size. If the core were to contract under the gravitational pressure of the star's outer envelopes, the temperature in the core would rise and the rate of nuclear reactions would increase. These reactions would produce more light—more photons. On their way out of the star, the photons would lose some of their energy to the gas atoms they interacted with, thus raising the gas pressure and causing the star to expand. This would lower the temperature in the core and slow the rate of nuclear reactions, leading to a fresh contraction.[3] The push and pull between gravity and gas pressure—regulated by the nuclear reactions in the core—causes stars to pulse. Like us, stars are governed by the beating of their hearts.

When molecular clouds break apart, each clump will be composed of a different amount (i.e., mass) of gas. Try breaking up a ball of wet sand and you will find that you are left with a few large clumps but many smaller ones. The same is true for stars: there are many more low-mass stars than high-mass stars.

How massive a star is at birth will determine everything about its life. More massive stars have hotter cores,

The push and pull between gravity and gas pressure—regulated by the nuclear reactions in the core—causes stars to pulse. Like us, stars are governed by the beating of their hearts.

which cause them to burn (via nuclear fusion) through their hydrogen stores faster than less-massive, cooler stars. The most massive stars burn through their hydrogen in several million years; the least massive stars were born billions of years ago and have tens of billions of years still to go.

The star's temperature will also determine its color (figure 12). Because of the way in which light makes its way out of the star (see note 2 in this chapter), most of that light will be concentrated into a narrow range of wavelengths; in hotter stars, the light will be concentrated at shorter wavelengths than in cooler stars.[4] When translated to color, this means that hotter stars will appear bluer than cooler stars. Our Sun, which isn't too hot and isn't too cold but just right, is yellow-green.

While the star continues to burn hydrogen in its core it is called a **main sequence** star. Once the hydrogen in the star's core is depleted, fusion will shut off and the temperature in the core will decrease. The cooling gas in the core will exert less and less outward pressure and the star will resume contracting under the weight of its own gravity.

Due to the contraction, the temperature in the core will once again start rising until it is high enough for hydrogen to start burning again in a layer of gas above the helium core. The core, meanwhile, will continue to contract until its temperature reaches the point where helium can begin fusing into carbon and oxygen. The star will

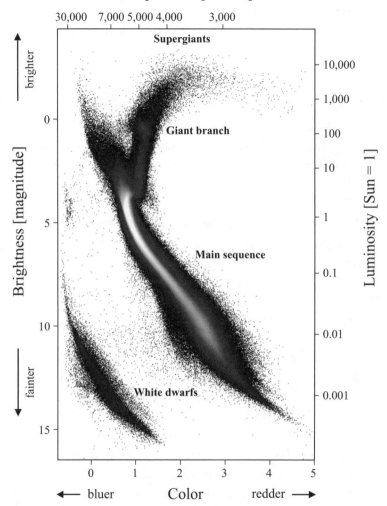

Figure 12 More luminous stars are hotter and bluer than dimmer stars. Stars begin their life burning hydrogen on the "main sequence," then become helium-burning giants. Stars less massive than eight times the mass of our Sun will become dim, cool white dwarfs, while more massive stars will become luminous, hot supergiants. Source: ESA/Gaia/DPAC, modified by the author.

now be composed of three layers: a helium-burning core, a hydrogen-burning layer above the core, and an inert layer of hydrogen above that.

The fusion processes in the helium- and hydrogen-burning layers will once again counteract gravity and the collapse will halt. Moreover, the star's outer layers will puff up, causing the temperature on its surface to go down. The star will be larger than before and, because of its lower surface temperature, appear redder. During this **red giant** phase, the star will continue to fuse helium into carbon and oxygen until the helium fuel in the core is depleted. The next stage in the star's life will depend on its initial mass and will decide its ultimate fate: death by ice or by fire.

Type Ia supernovae

In stars with masses up to eight times the mass of our Sun, the sequence of core contractions will stop at the carbon-oxygen phase. This time, when the helium fuel in the core is depleted, the ensuing contraction will not raise the temperature in the core high enough to ignite carbon and begin the next fusion stage. Instead, the core's contraction will be halted by a quantum-physics phenomenon called electron **degeneracy pressure**: when electrons are squeezed together at high pressures, they begin to press back.[5]

If left alone, white dwarfs will cool down, like embers, for the rest of time. This will be the final fate of our Sun.

The star's carbon-oxygen core will be small (about the size of Earth) and white-hot, earning it the appellation **white dwarf**. Given its high temperature, the white dwarf will mostly emit ultraviolet photons, which will be energetic enough to interact with the outer helium and hydrogen layers of the star and push them out into space in a gentle "wind." Lit by the radiation from the white dwarf, the expanding envelope will glow for a few thousand years until the expanding layers of gas become thin enough for the white dwarf's light to pass through undisturbed. This short but beautiful phase is called a **planetary nebula**.[6]

If left alone, white dwarfs will cool down, like embers, for the rest of time. This will be the final fate of our Sun. Yet, because of various hints gleaned from their light curves and spectra, it is now widely accepted that **Type Ia supernovae** are thermonuclear explosions of carbon-oxygen white dwarfs. The question, then, is how to provoke this type of star to explode. As we will see in chapter 7, the trick may be to place the otherwise stable white dwarf in an unstable relationship with another star.

Core-collapse supernovae

In stars with masses in the range of 6–12 Solar masses, the carbon-oxygen core will go through another contraction until the carbon is ignited and another fusion stage sets

Type Ia supernovae
are thermonuclear
explosions of carbon-
oxygen white dwarfs.

in. The star will now have several layers: a carbon-burning core surrounded by a helium-burning shell, a hydrogen-burning shell, and an inert hydrogen shell. Stars in this phase are larger and more luminous than red giants, and are called "supergiants." The fate of these stars, which is still uncertain, is presented in chapter 7.

In stars more massive than 12 Solar masses (i.e., at least twelve times more massive than our Sun), once the carbon is depleted, another contraction will ignite neon. The next contraction will ignite oxygen, and then silicon. Each new fusion phase is shorter than the last. For example, a 25-Solar-mass star will spend ~7 million years fusing hydrogen into helium, ~800,000 years burning helium, another ~500 years burning carbon, roughly one year burning neon, five months burning oxygen, and less than a day burning silicon.[7] Finally, the silicon-burning stage leaves a core composed almost completely of iron.

So far, the fusion of light elements into heavier ones has resulted in a release of energy in the form of photons and neutrinos (elementary particles with no electric charge and small—but as yet undetermined—masses). Iron is the last element in this chain. In order to fuse iron into heavier elements, external energy must be injected into the fusion process. As no such energy source exists, the next few seconds will be the star's last.

Instead of igniting a new fusion stage and stabilizing, the temperature and density in the iron core will continue to rise until several nuclear processes working in tandem will turn the star's core into a nucleus composed almost solely of neutrons (electrically neutral particles that, together with protons, form the nuclei of atoms).[8] This "neutronization" of the core conspires to remove energy and pressure support from the core and triggers a fatal collapse.

In stars with an initial mass of less than 25 Solar masses, the core will continue to collapse until neutron degeneracy pressure kicks in and counteracts gravity (see note 5 in this chapter). The newborn **neutron star** ceases its collapse with a slight bounce, which sends a shockwave into the still-collapsing outer layers of the star. Together with the stream of energy-bearing neutrinos released by the neutronization of the core, the shockwave rips into the outer layers of the star and blows them into space in a supernova explosion. Type II, Ib, and Ic supernovae are all thought to explode via this process and are collectively called **core-collapse supernovae**.

In stars more massive than ~25 Solar masses, but less massive than 40 Solar masses, the shock released by the formation of the neutron star may not be strong enough to overcome the pressure of the rest of the star still collapsing onto the core. The infalling matter will grow the neutron star's mass until neutron degeneracy pressure will no longer be enough to resist gravity, and the neutron star will collapse into a **black hole**. The predicted evolution of stars more massive than 40 Solar masses, which

Type II, Ib, and Ic supernovae are all thought to explode via core collapse.

may result in "pair-instability supernovae," is covered in chapter 7.

Supernova remnants

Although it takes mere seconds for a star to explode, the effects of the ensuing supernova will be felt for thousands of years.

First, a stream of neutrinos, created by the birth of a neutron star or through the many nuclear processes triggered by the explosion, will rush out into space, barely interacting with anything in its path. If the star is somewhere in our own Milky Way Galaxy, specialized neutrino telescopes might manage to detect a handful of these elusive particles (chapter 8).

Next, light produced by the radioactive decay of various elements created during the explosion will filter out through the blown-out layers of the star. Some of that light will make its way to our telescopes, and the supernova will be discovered (plate 2). Depending on the telescope used to observe the supernova, along with the supernova's type, distance, and intrinsic brightness, the light curves described in chapter 3 will be visible across the electromagnetic spectrum (from X-rays and ultraviolet, through the optical and near-infrared, to radio) for days, months, or even years. The light curve of SN 1987A (figure 9), the closest supernova to have exploded in modern times, is still regularly observed by astronomers.

Once the light created by the explosion fades away, the effects of the supernova on its immediate surroundings will begin to emerge. Plate 3 shows the **supernova remnants** left over from the five confirmed historical supernovae (chapter 1). Not to be confused with the stellar remnant (the neutron star or black hole), the supernova remnant is a nebula of glowing gas formed by the supernova ejecta (the blown-up gas that used to be the star) as they slam into the gas that surrounded the star before it exploded.[9]

Stars, including our Sun, lose mass on a regular basis through stellar winds pushed out either by the pressure of the hot gas on the surface of the star or by the interaction of ultraviolet radiation with the hydrogen in the outer layers of the star. Massive stars are also thought to undergo a series of eruptions before they finally explode (the smaller peak in the Type IIn light curve shown in figure 9 is considered to be such an eruption). Depending on the initial mass of the star, these winds and episodic eruptions can remove a large fraction of the star's mass. Beyond the gas expelled by the star itself lies the low-density gas that populates the space between stars.[10]

The interaction of the supernova ejecta with the gas surrounding the star creates a shockwave (called the "forward shock") that accelerates the gas and heats it to

Although it takes mere seconds for a star to explode, the effects of the ensuing supernova will be felt for thousands of years.

temperatures that cause it to glow in X-rays. The supernova ejecta, in turn, are sapped of energy and begin to slow and cool. At the same time, a second shockwave (called the "reverse shock") forms and begins to reheat the ejecta behind the expanding blast wave. Electrons accelerated by the magnetic field along the forward shock emit light at radio wavelengths. This is the main source of light from a supernova remnant; more than 90% of the 294 supernova remnants discovered so far in our Milky Way Galaxy were detected with radio telescopes.[11]

A schematic description of the supernova remnant at this ejecta-driven stage is shown in figure 13. At this stage, the shape of the remnant depends on the density structure of both the ambient gas and the supernova ejecta. Type Ia supernova remnants, which are thought to explode in environments with little surrounding gas, are expected to be symmetric and round. A good example is Tycho, the remnant left over from SN 1572, shown in plate 3. Core-collapse supernovae, which explode in star-forming regions rich with gas, are expected to create more asymmetric remnants, such as 3C 58 and the Crab Nebula.

Hundreds to thousands of years later, the forward shock begins to slow down and cool, leading to the formation of "dimples" and curling "fingers" where the inner layer of supernova ejecta comes into contact with the outer layer of swept-up, shocked gas. The Tycho supernova remnant is a good example of this stage.[12]

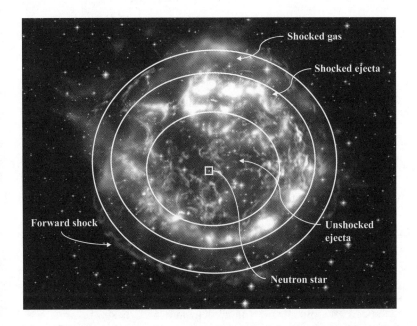

Figure 13 X-ray image of supernova remnant Cassiopeia A, a Type IIb supernova that exploded in the seventeenth century CE. Overlaid is a schematic structure of the initial phase of supernova remnants. As the supernova ejecta expand into the ambient gas surrounding the star's location, the gas is heated and accelerated by the forward shock. At the same time, an inward-facing reverse shock heats the oncoming unshocked supernova ejecta. The neutron star created by the supernova is highlighted at the center of the remnant. Credit: NASA/JPL-Caltech/STScI/CXC/SAO; contrast enhancement and schematic by the author.

For the first few thousand years, the hot supernova remnant only radiates X-rays and radio waves, but as the forward shock slows down, the shocked material cools and begins to radiate in other wavelengths as well—ultraviolet, optical, and near-infrared. Radiating at these wavelengths is more efficient than in X-rays; it removes more energy from the shocked material (thus "cooling" it) and decelerates the shock even further.

Remnants at this stage have been interacting with interstellar gas of varying densities for thousands of years. No matter how round the remnants might have been to begin with, their shapes will now be asymmetric and often disjointed.

Finally, after a few tens of thousands of years, the shockwave will slow down to a velocity lower than the speed of sound and the shockwave will dissipate and fade away.[13]

In core-collapse supernovae, the supernova remnant will also contain a stellar remnant—a neutron star or black hole—left behind by the dying star. As the light from the explosion fades away, the neutron star becomes visible in radio and X-ray wavelengths in the form of a **pulsar**.[14]

Most stars, including our Sun, rotate around a central axis. Ice skaters with outstretched arms spin faster once they fold their arms close to their body. Likewise, as the core of a star contracts to a white dwarf, a neutron star, or a black hole, its spin will also increase. Spinning neutron stars—pulsars—have periods ranging from milliseconds to tens of seconds. They also have strong magnetic fields, which channel charged particles into the poles, leading to the projection of beams of light. If oriented correctly relative to Earth, these beams will pass across a telescope's field of view periodically and the neutron star will seem to pulse, like a lighthouse.

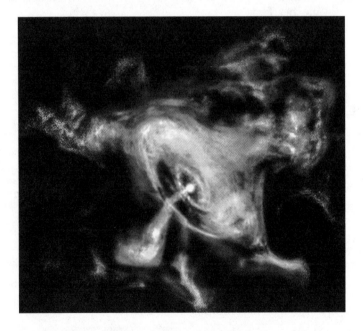

Figure 14 X-ray image of the center of the Crab Nebula, showing the spinning neutron star (the bright spot in the center of the image) surrounded by a spinning pulsar-wind nebula and shooting off a polar beam of light. Source: CXC.

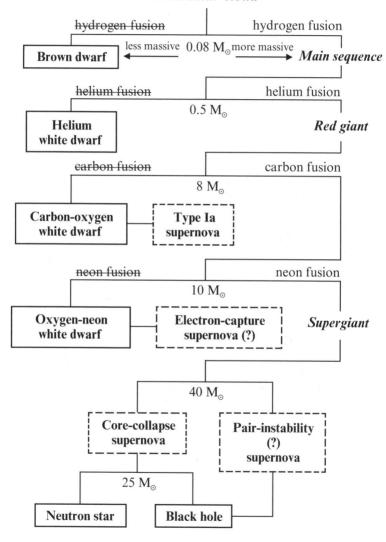

In plate 3, the neutron stars created by the supernovae of 1054 and 1181 CE are the bright white objects at the centers of the Crab Nebula and 3C 58, respectively. The jet and spinning clouds of charged particles that make up the **pulsar-wind nebula** around the spinning neutron star at the center of the Crab Nebula are shown in figure 14 and recreated in color in plate 3.

The lives and deaths of stars, summarized in figure 15, strongly depend on how massive they are when they are born. Stars live millions to billions of years, but the explosions that tear them apart last mere seconds. The ensuing supernovae remain visible for a few tens of thousands of years—a blink of an eye in cosmic time—and are then extinguished forever. Yet through their death, some stars give birth to new types of objects—neutron stars and black holes—with complex and exciting physics of their own. In the next chapter, I describe the crucial role played by supernovae in our own existence and how one day they might put an end to that as well.

Figure 15 A star's fate depends most of all on its initial mass, M. Below ~0.08 Solar masses (denoted by the symbol $M_\odot$), the temperature in the core is too low to ignite hydrogen fusion. Above that threshold, stars with different initial masses will go through divergent evolutionary stages (italics) and end up as various types of stellar remnants (solid boxes). Supernova explosions are shown in dashed boxes. At each bifurcation, less massive stars fall to the left while more massive stars fall to the right. Electron-capture and pair-instability supernovae are discussed in chapter 7.

SUPERNOVAE, THE UNIVERSE, AND US

Supernovae do not explode in a vacuum. While space is often described as an empty, never-changing vacuum, that is only because we view it through human eyes. On the scales we are used to—kilometers and years—space would indeed appear to be empty and static. But on cosmic scales—parsecs and billions of years—space is littered with gas, dust, stars, and planets. Supernovae interact with all of these objects and help shape our Galaxy, our planet, our bodies, and perhaps, one day in the distant future, our extinction.

Life-givers: creating and dispersing the elements that make up our world

The kind of matter we experience in our daily lives makes up less than 5% of the Universe. **Dark matter**, an invisible, still unknown form of matter that only interacts with normal matter through gravity, makes up a quarter of the Universe's mass. The rest, roughly ~70%, is in the form of the even more mysterious **dark energy**. All we know about the latter is that it is somehow accelerating the expansion of the Universe by counteracting gravity on cosmological scales (chapter 6). "Regular" matter—usually called **baryonic matter** by astrophysicists—is made up of atoms of various elements, in turn composed of protons, neutrons, electrons, and so on. Supernovae play a key role in both the creation of the elements and their dispersal through space.

The nucleus of hydrogen, the simplest element, is composed of a single proton. The next element in the periodic table—helium—is composed of two protons and two neutrons. Together, these two elements account for almost all of the regular matter in the Universe: ~74% hydrogen and ~24% helium. All the elements heavier than helium, from lithium to uranium, account for less than 2% of the visible Universe. Collectively referred to as "metals" by astronomers, these are the elements that make up everything we see and touch, from the air that we breathe (mostly nitrogen and oxygen), through the salt we use to garnish our food (sodium and chlorine), to the aluminum in airplane fuselages and the foil we use to wrap up leftovers.

Hydrogen, used in rocket fuel, was created during the **Big Bang**. So was most of the Universe's helium, often used

The kind of matter we experience in our daily lives makes up less than 5% of the Universe.

to inflate party balloons and make our voices sound comically high.[1] Some lithium (used in mobile phone batteries), and all boron (found in both soap and fireworks) and beryllium (a light and durable—but expensive—metal used in the production of, for example, gyroscopes, missiles, and golf clubs), were created through a nuclear process in which highly energetic charged particles bombard an atom and cause it to split into atoms of lighter elements.[2] All of the other elements in the periodic table are continuously being created by stars and their explosive deaths (plate 4).[3]

The chain of element creation—**nucleosynthesis**—starts with the fusion of hydrogen into helium during the first stage of a star's life.[4] During the next, red giant stage, the star's core is dense and hot enough that three helium atoms can fuse together to form stable carbon. Next, oxygen is created by fusing carbon and helium. Helium atoms are also referred to as *alpha* (α) particles, and anytime an atom of a given element fuses with a helium atom to create an atom of a heavier element, it is said to undergo the alpha process.

As the star progresses to carbon burning, the temperature and density in the core are high enough that heavier elements are created through a complicated mixture of nuclear reactions: capture and release of helium nuclei, electrons, positrons (the electron's antiparticle), neutrons, or photons. An element produced by one nuclear reaction may be radioactively unstable and decay, via a different nuclear reaction, into another element. For example, two carbon atoms can fuse into an unstable form of magnesium, which can then decay into stable magnesium (plus a neutron), neon (along with a helium atom), or sodium (and a proton).

The neutrons, protons, and helium atoms released by these reactions can go on to react with the new elements. Some reactions are more common than others, so that each burning stage results in principal products (carbon turns into magnesium and neon) and secondary ones (carbon burning also creates some silicon and phosphorus). In this manner, the final burning stages in the star's core see to the creation of all elements up to the iron-group elements from titanium to zinc.

Elements heavier than zinc are created through two mechanisms dubbed the *s* and *r* processes. In both processes, the nucleus of a seed element absorbs a neutron from the surrounding gas. In the *s* process (where *s* stands for *slow*), the neutron has enough time to decay into a proton. The new nucleus now has an extra proton, which turns it into the nucleus of the next element up the periodic table. In the *r* process (where the *r* stands for *rapid*), a second neutron, or more, is absorbed by the original atom before the first neutron has had enough time to decay. If the original nuclei are embedded in a neutron-rich environment, they will rapidly absorb large numbers of neutrons and create a range of heavier isotopes of the same element. The nature of an element is determined by the number of protons in its atom. Carbon, for example, will always have six protons. But it can have more (or fewer) than six neutrons. Isotopes are versions of an element with the same number of protons but a different number of neutrons.

Most of the new elements created via these two processes will be radioactively unstable and, over time, will decay into lighter elements. These will decay in turn until stable isotopes are finally reached. Most of the s-process elements are thought to be produced in the helium-burning stage of massive stars or during flashes of burning in the helium shells of lower-mass stars. The r process, however, requires the kind of neutron density only thought to be found in supernovae and the mergers of neutron stars (chapter 8).

Some elements are created via "explosive nucleosynthesis" during supernova explosions. In core-collapse supernovae, the blast wave traveling outward from the core raises the temperature in the outer shells of the star to the point where they are ignited. Through the same set of nuclear reactions (fusion, neutron or electron capture and release, and photodisintegration through the absorption of photons), all of the elements up to the iron group are created.[5]

In Type Ia supernovae, the high-temperature ignition of carbon in the white dwarf is thought to lead first to a "deflagration" phase during which the blast wave travels through the white dwarf at subsonic velocities, i.e., below the speed of sound. This relatively low-density burning stage produces elements up to silicon. The blast wave is then thought to transition to a second, "detonation" phase during which it moves through the white dwarf at supersonic velocities (higher than the speed of sound), and the density is high enough to create the iron-group elements.[6]

Without supernovae, most of the elements in the Universe would remain locked inside the stars that create them. Most stars, including our Sun, experience stellar winds that, every year, eject a small fraction of their mass into space. However, these winds only become the dominant mode of mixing new elements into space in stars more massive than 20 Solar masses. Even then, mass loss through winds occurs mostly during the hydrogen- and helium-burning stages of a star's life, which result in the creation of elements lighter than silicon.[7] All of the other elements are expelled into space by supernovae and mixed into the space between the stars by the interaction of the supernova remnant with its surrounding gas.

Eventually, new stars will be born out of this enriched gas and will begin their own chain of element creation. Each new generation of stars starts out with a higher fraction of elements heavier than hydrogen; astronomers refer to this fraction as "metallicity."

Without supernovae, most of the elements in the Universe would remain locked inside the stars that create them.

Star formation and galactic winds

Beyond enriching successive generations of stars, super-novae are also thought to play a role in regulating the rates at which stars are formed. Observations of new stars being formed close to the edges of supernova remnants originally led astrophysicists to presume that the expand-ing remnants were condensing gas at their outskirts into molecular clouds that could then collapse and form stars.

However, research over the last couple of decades has cast doubt on this picture. As the supernova remnant expands into space, it will not only condense the gas it encounters but heat it as well. For molecular clouds to collapse into stars, the gas inside them needs to be cold. As in stars, hot gas will press outward, resisting gravity and stalling the collapse of the molecular cloud. Cold gas, made up of slow-moving particles, will have a harder time resisting the inward pull of gravity. It is possible that, as the remnant's shock decelerates, it will slow down to a ve-locity high enough that the shock can still condense the gas in front of it but low enough so as not to heat the gas considerably.[8]

A different way for supernovae to affect star forma-tion is through the creation of **superbubbles** (plate 5). Because stars are formed out of the collapse of molecular clouds, they are usually found in clusters of hundreds of stars. The strong winds from the more massive, hot stars in these clusters will carve out cavities around the stars. These young, hot stars will also be the first to explode as supernovae and will do so in rapid succession (within a few million years of each other). The shocks of their supernova remnants will coalesce with each other and with the earlier winds and excavate a larger cavity, hundreds of light-years wide.

As with supernova remnants, once the expanding edges of a superbubble slow down to a few tens of kilometers per second, they may be able to compress the gas ahead of them into molecular clouds that will then collapse to form stars. The shell at the edge of the superbubble is also expected to eventually succumb to gravity and fragment into dense clumps that may themselves become star-forming regions. Such star-forming regions, distributed in shell-like configurations centered on clusters of young stars, have been observed in the Milky Way and its satellite galaxies.[9]

In galaxies with extremely high rates of star formation, known as "starburst" galaxies, the superbubbles formed by winds and supernovae can grow to such dimensions and temperatures that, instead of forming shell-like structures at their edges, they will continue to expand and heat the surrounding gas. The energy injected into the hot gas by the winds and supernova remnants is enough for the gas to punch its way out of the galaxy in **galactic winds** or outflows (plate 6).

Expelling hot gas into the space surrounding the galaxy creates a feedback loop that regulates the rate at which stars can form.[10] High rates of star formation result in more young supernovae, which then drive strong outflows out of the galaxy. The removal of gas from the galaxy means fewer stars can be created and the star-formation rate goes down. The hot gas expelled from the galaxy will cool and eventually rain back into the galaxy, where it will once again clump into molecular clouds and form new stars. Because the gas was expelled by interaction with supernova remnants, it will be enriched with the elements produced and dispersed by the supernovae, so that the new wave of stars created by the infalling gas will be enriched as well.[11]

Cosmic-ray accelerators

As physicists in the nineteenth century began to study radioactivity, they discovered a background source of radiation that interfered with their experiments. At first, it was assumed that the source of this radioactivity was either the Earth's crust, its atmosphere, or the Sun.[12]

Throughout 1912, Victor Hess undertook seven hot-air balloon flights during which he measured the intensity of this background radiation as he ascended through the air. As he rose in his balloon, the radiation's intensity would first drop, then begin to increase steadily, ruling out the Earth's crust as the radiation's source. Some of Hess's flights were conducted at night or during Solar eclipses, allowing him to rule out the Sun as another source for the radiation. He concluded that the radiation was of extraterrestrial origin.[13]

Later dubbed **cosmic rays**, this radiation is composed of electrically charged particles: mostly protons, but also nuclei of helium and heavier elements. Measurements of the energies of these particles reveal that they travel through the Galaxy at velocities close to the speed of light. To reach such high velocities, it is thought that these particles are swept up by the forward shocks of supernova remnants and then accelerated by repeatedly passing back and forth through the shock and interacting with the magnetic turbulence of the hot gas on either side.[14]

Cosmic rays are a major source of danger for astronauts. A cosmic ray hitting a human cell could lead to mutations and, eventually, cancer.[15] Pilots and airplane crews, as well as anyone living at high elevations, are also at risk, though vastly less than astronauts. Not to worry, though: most of the cosmic rays hurled at the Earth by supernova remnants never reach the surface. They interact with particles in the Earth's atmosphere, where they break apart into a cascade of particles with lower and lower energies. This cascade also results in a shower of blue light called Cherenkov radiation, which is picked up by specialized

SN 2018gv

Plate 1 SN 2018gv, the bright blue starlike object in the bottom right of the image, is several weeks past its peak yet still one of the brightest objects in its host galaxy. Source: Data from O. Graur et al., "A Year-long Plateau in the Late-Time Near-Infrared Light Curves of Type Ia Supernovae," *Nature Astronomy* 4 (February 2020): 188–195.

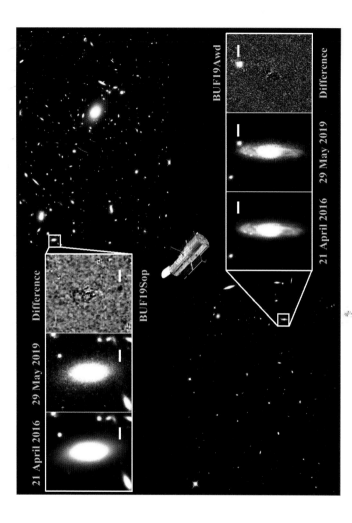

Plate 2 Two supernovae discovered at the same time with two different cameras on the Hubble Space Telescope using the difference-imaging technique. Credit: O. Graur (U. Portsmouth), A. M. Koekemoer (STScI), C. L. Steinhardt (U. Copenhagen), and M. Jauzac (Durham U.). Image of the Hubble Space Telescope: NASA/ESA.

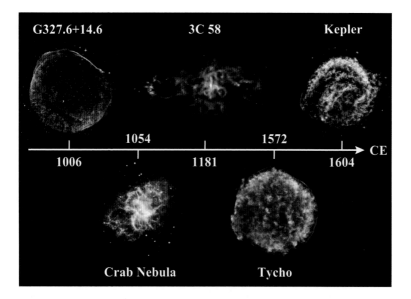

Plate 3 Remnants of the five historical supernovae that exploded in 1006 (G327.6+14.6), 1054 (Crab Nebula), 1181 (3C 58), 1572 (Tycho), and 1604 (Kepler) CE. Sources: **G327.6+14.6:** X-ray: NASA/CXC/Rutgers/G. Cassam-Chenaï, J. Hughes et al.; radio: NRAO/AUI/NSF/GBT/VLA/Dyer, Maddalena & Cornwell; optical: Middlebury College/F. Winkler, NOAO/AURA/NSF/ CTIO Schmidt & DSS. **Crab Nebula:** NASA, ESA, G. Dubner (IAFE, CONICET-University of Buenos Aires) et al.; A. Loll et al.; T. Temim et al.; F. Seward et al.; radio: VLA/NRAO/AUI/NSF; X-ray: Chandra/CXC; Infrared: Spitzer/JPL-Caltech; ultraviolet: XMM-Newton/ESA; optical: Hubble/STScI. **3C 58:** X-ray: NASA/CXC/SAO. **Tycho**: X-ray: NASA/CXC/Rutgers/J. Warren & J. Hughes et al. **Kepler**: X-ray: NASA/CXC/NCSU/S. Reynolds et al. Timeline added by the author.

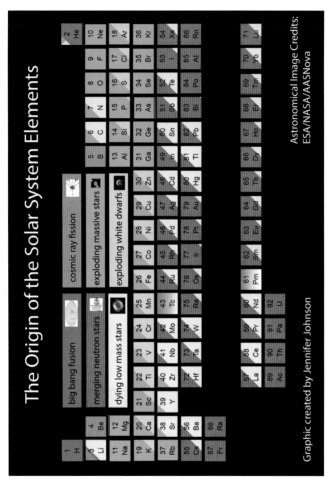

Plate 4 Periodic table of the elements and the processes by which each element is created. Some elements are created by more than one process, as indicated by the different color combinations. Elements heavier than uranium are either unstable or decay quickly and so are not shown here. The same is true of technetium (Tc) and promethium (Pm), shown here in gray. Source: Jennifer Johnson.

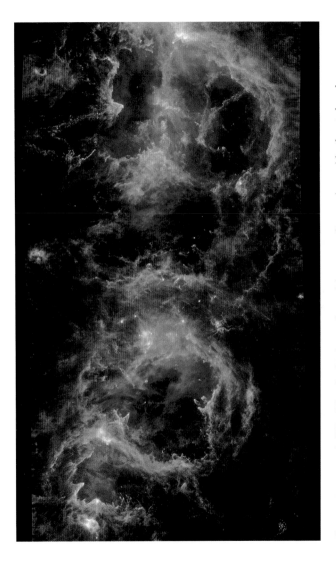

Plate 5 Three superbubbles (blueish-white) carved out of the surrounding gas (orange) by the winds and supernova explosions of young, massive stars. Credit: ESA/Herschel/NASA/JPL-Caltech; acknowledgement: R. Hurt (JPL-Caltech).

Plate 6 Messier 82 (or M82), a nearby starburst galaxy, displays strong outflows of warm hydrogen (red) driven by the winds and supernova explosions of young, massive stars. Source: NASA, ESA and the Hubble Heritage Team (STScI/AURA). Acknowledgment: J. Gallagher (University of Wisconsin), M. Mountain (STScI) and P. Puxley (NSF).

Plate 7 SN Refsdal, a multiply imaged, strongly lensed supernova, appears as four yellow stars in an Einstein cross configuration. Each of the images shows the supernova as it was at a different point in time. The supernova reappeared a year later in a separate image of its host galaxy. Source: NASA, ESA, S. Rodney (Johns Hopkins University, USA) and the FrontierSN team; T. Treu (University of California Los Angeles, USA), P. Kelly (University of California Berkeley, USA) and the GLASS team; J. Lotz (STScI) and the Frontier Fields team; M. Postman (STScI) and the CLASH team; and Z. Levay (STScI).

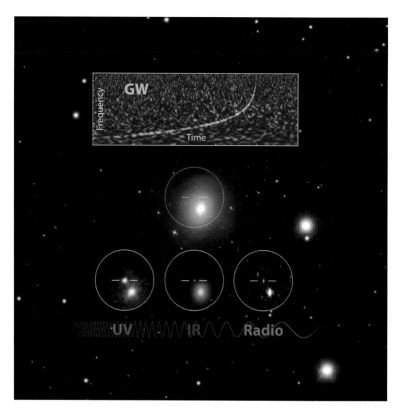

Plate 8 The gravitational-wave event GW170817 and its associated kilonova. The top panel shows the gravitational-wave signal detected by the Laser Interferometer Gravitational-Wave Observatory (LIGO): a quick rise in frequency of the gravitational waves emitted by the neutron stars as they spiral in and merge. In the background is a pre-explosion Hubble Space Telescope optical image of the host galaxy of the event. The kilonova, as imaged in the ultraviolet (UV), infrared (IR), and radio, appears in the three bottom insets. Credit: courtesy of Robert Hurt (Caltech/IPAC), Mansi Kasliwal (Caltech), Greg Hallinan (Caltech), Phil Evans (NASA), and the GROWTH collaboration. For sources, see note 14 of chapter 8.

Cherenkov array telescopes used to study the origins and physics of cosmic rays.

Life-takers: supernovae and mass extinctions

We owe our existence to supernovae; without them, there would be no calcium for our bones or iron for our blood. Yet supernovae may one day be the death of us. As the Sun moves through its roughly 230-million-year orbit around the center of the Milky Way Galaxy, it passes through environments with differing densities of stars and gas. It is currently traveling through a region of low-density material called the Local Bubble, which is thought to have been excavated within the last 10 million years by several supernovae of stars located in the Scorpius-Centaurus association of young stars.[16] The last decade has seen mounting evidence that some of these supernovae may have left traces of their existence on the Solar System, including on Earth and the Moon.

The blast wave from a nearby supernova may explain the compact size of the Solar System (40–60 AU), relative to the sizes of young planetary systems observed in our Galaxy (100–400 AU), and why the equator of the Sun is tilted relative to the plane of the planets in its orbit.[17]

Down on Earth, recent searches for supernova signatures have focused on one of the radioactive isotopes

We owe our existence to supernovae; without them, there would be no calcium for our bones or iron for our blood. Yet supernovae may one day be the death of us.

of iron created during supernova explosions, ^{60}Fe (pro-nounced "iron-sixty"). With a half-life of ~2.6 million years,[18] any trace of ^{60}Fe incorporated into the Earth when it was formed ~4.5 billion years ago will be gone by now. Yet several studies of samples of ferromanganese crust, deep-sea sediment, and microfossils of metal-reducing bacteria extracted from the Earth's oceans have revealed strong ^{60}Fe signals originating some 2 million years ago.[19] An analysis of Lunar soil samples collected by the Apollo 12, 15, and 16 missions also revealed an excess of ^{60}Fe that had been deposited on the Moon 1.7–2.6 million years ago.[20] Finally, several puzzling features in the local cosmic-ray energy spectrum can be explained by an injection of cosmic rays from a nearby source approximately 2 million years ago.[21]

The amount of ^{60}Fe and the total energy in cosmic rays required to explain the signals measured by these various studies point to several supernovae occurring nearby and impacting Earth ~2 million years ago. This time frame co-incides with a mass extinction event of marine life that occurred during the Pliocene-Pleistocene boundary ~2.6 million years ago, which wiped out a third of all large ma-rine animals.[22] Whether or not this extinction event was caused by nearby supernovae or other factors, such as re-duction of coastal habitats due to changes in sea levels,[23] is still under debate. Radiation from a supernova could lead to ozone depletion, which would weaken the atmosphere's ability to block ultraviolet rays that are harmful to life. However, the supernova would have to be close by (<10 parsecs), much closer than the probable origin of the supernovae in the Scorpius-Centaurus association (40–130 parsecs).[24]

As we learn more about the possible impact of nearby supernovae on Earth in the past, we must also prepare for any such impact in the future. Luckily, several factors conspire to keep us safe, at least for the next few million years: the vastness of space, the close proximity required for a supernova to have any harmful effect on terrestrial life, and the million- to billion-year time scales of the Sun's motion and the life cycles of the stars it encounters.

Currently, the closest star that might explode as a supernova is IK Pegasi, a two-star (binary) system that includes a white dwarf, which may, billions of years from now, explode as a Type Ia supernova. Next is Betelgeuse, a red supergiant easily visible to the naked eye as part of the Orion constellation. Betelgeuse will explode as a core-collapse supernova anytime in the next million years. At a distance of 200 parsecs, its explosion will light up the sky but be too far away to have any harmful effect on us. The same is true for Eta Carinae, a massive star that has already undergone several eruptions, a sign that it too may be close to explosion. Yet it is even farther away than Betelgeuse—2,300 parsecs—and so holds no threat for Earth.

As the Sun moves throughout the Milky Way Galaxy, the distances between it and various supernova progenitors will change over time. Taking into account the frequency of supernovae in our Galaxy, a simulation of the Sun's future orbit and the evolution of the stars it will pass by shows that in the next 2 billion years, the Solar System could be impacted by ~20 supernovae closer than 10 parsecs.[25] So, while it is quite possible that a nearby supernova will, one day, cause a mass extinction event on Earth, we are probably safe for millions of years. In the meantime, we should be far more worried about other space-borne dangers, such as asteroid impacts, or human-caused catastrophes such as climate change. Unfortunately, we cannot use supernovae to solve our problems. But, as we will see in the next chapter, we can use them to study other astrophysical phenomena, as well as the very makings of the Universe.

SUPERNOVAE AS TOOLS

Draped in long, white lab coats and protected by clear plastic goggles, scientists are often pictured leaning over bubbling beakers or an optical table alight with zigzagging red lasers. In these gleaming laboratories, biologists, chemists, and physicists can control temperature, humidity, acidity, seismic interference, and anything else their hearts desire. They can cross different fly strains, design new bacteria, move atoms one by one, and even freeze light. They can use off-the-shelf instruments or design and fabricate their own. They are the masters of their experiments.

Astrophysicists have observatories filled with telescopes and cameras, but these are not our laboratories. We design and build cutting-edge cameras, but we cannot use them to construct a white dwarf or blow up a red giant. For astronomers, the Universe is our lab. We have to rely on the objects and the phenomena available to us in the night sky. But, never content to be left at the mercy of the Universe, wherever we can we turn the very phenomena we study into tools. Below, I describe a few of the main ways in which supernovae are used as tools to study other astrophysical phenomena, basic physics, and the very foundations of the Universe.[1]

Supernovae as cosmological rulers

Fans of the British comedy troupe *Monty Python*, along with anyone who grew up watching *Animaniacs* cartoons, will happily tell you—bursting into song—that we live on a planet that revolves around a star, which is one of billions in a galaxy called the Milky Way, which itself is one of millions and billions in an amazing and expanding Universe.[2] Yet the realization that the Universe is larger than just our Galaxy—and that it is expanding—is merely a century old. And as we learned some twenty years ago, for reasons we do not fully understand yet, the Universe's expansion is accelerating.

The last realization is due to the use of Type Ia supernovae as a type of ruler to measure the size of the Universe. It is one thing to describe the Universe as containing billions of galaxies, but how big is it, in meters, yards, or light-years? The first ruler you ever used was probably a simple piece of wood or plastic with notches along its longer edge.

For astronomers, the Universe is our lab.

Each notch corresponded to an agreed-upon unit of distance: a centimeter or an inch. It was good enough for measuring the lengths of lines in math problems at school, but too short for everyday uses such as measuring your height or the size of your bedroom. For that, you probably turned to a different kind of ruler—a rollup tape measure.

To measure distances on astronomical scales we need new kinds of rulers and tape measures. The distance to the Moon (and the rate at which it is moving away from us) is measured by aiming powerful lasers at reflective panels left on the Lunar surface by Apollo astronauts and measuring the time it takes the laser pulses to go back and forth. Distances to asteroids in the asteroid belt beyond Mars are measured using radar. Distances to the nearest stars are measured using their parallax: a small shift in their perceived position against the backdrop of more distant, seemingly static stars caused by the Earth's orbit around the Sun (similar to the effect of holding your thumb out in front of your nose and then closing each eye in succession; your thumb will appear to move against the background).

The units used to measure distance also change with distance. The distance to the Moon is still measured in meters, but distances to other planets are measured in astronomical units (AU), defined as the distance between the Earth and the Sun. Distances to other stars are measured in light-years—the distance light travels in a single year—or parsecs, where 1 parsec = 3.26 light-years.

Distances across our Galaxy are measured in thousands of parsecs (kiloparsecs), to nearby galaxies in millions of parsecs (megaparsecs).

Beyond the nearest stars, astronomers use "standard candles"—objects whose luminosity behaves in well-understood ways—to construct a distance ladder. Imagine holding two candles, both burning at the same intensity. Place one candle in the corner of a room and start walking away from it. The farther you walk, the fainter the candle will seem when compared to the second candle still in your hand. By comparing the brightness of the two candles, and taking into account that the brightness of a light source diminishes with the square of the distance from it, you will be able to measure the distance between the two standard candles. Now place the two candles at different distances from you. Measure the distance to the closer candle with a ruler. Add to that the estimate of the distance between the candles that you derived before and you have a measurement of the distance between you and the farther candle. The ruler and the candles were the first and second rungs on your distance ladder.

Distances to the nearest galaxies are measured by studying certain types of stars whose luminosity pulses—brightens and dims—in a periodic way. In 1912, Henrietta Swan Leavitt discovered that the more luminous of these variable stars (called Delta Cepheids) had longer periods than their less luminous counterparts.[3] Variable stars

Beyond the nearest stars, astronomers use standard candles—objects whose luminosity behaves in well-understood ways—to construct a distance ladder.

lying farther away will appear fainter but their periods of pulsation will remain the same. The Leavitt relation between the star's period and luminosity then allows us to compare its apparent brightness to its intrinsic luminosity and derive its distance. This is the first rung on the cosmological distance ladder.

Type Ia supernovae provide the next rung on the ladder. The large amount of radioactive nickel created by these supernovae makes them exceedingly bright; their luminosity can rival the combined light of the billions of stars that make up their host galaxies. They are also millions of times more luminous than pulsating variable stars; while the latter can only be observed in nearby galaxies, Type Ia supernovae can be seen clear across the Universe.

Type Ia supernovae are not exactly standard candles; some Type Ia supernovae are intrinsically more luminous than others. Yet, like the pulsating variable stars studied by Henrietta Leavitt, Type Ia supernovae are standardizable. During the 1970s, Bert Woodward Rust and Yury Pavlovich Pskovskii independently discovered that more luminous Type Ia supernovae had broader light curves; they took longer to reach peak and decline.[4] In 1993, Mark Phillips used well-sampled light curves taken with modern CCD cameras to modernize this width-luminosity correlation.[5] The Rust-Pskovskii-Phillips relation (figure 16) allowed astronomers to standardize Type Ia supernovae and compare supernovae discovered in galaxies with known

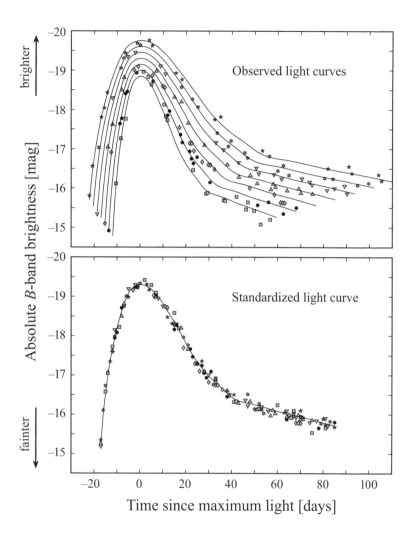

distances (e.g., galaxies with known pulsating variable stars) with fainter Type Ia supernovae that exploded in more distant galaxies.[6] The resultant distance, which depends on the brightness of the supernova, is called the "luminosity distance."

Other supernovae, such as Type IIP supernovae and superluminous supernovae (chapter 7), can also be standardized and used for cosmology, but the distances they produce are less precise than those measured with Type Ia supernovae and so are not in common use.[7]

The accelerating expansion of the Universe and dark energy

The use of supernovae to measure distances is further complicated by the expansion of the Universe. The Universe is not expanding *into* anything. By definition, the Universe is all there is. By expansion, astrophysicists mean that

Figure 16   Type Ia supernovae, used to measure cosmological distances, are not homogeneous: more luminous Type Ia supernovae have broader light curves than less luminous explosions. The relation between the luminosity of the supernova and the shape of its light curve is used to standardize observed supernovae and measure their distances. These depend not only on the actual distance between us and the supernovae but also on the expansion of the Universe, allowing astronomers to use Type Ia supernovae to characterize the expansion.

the distance between any two points in the Universe is growing larger over time. At small scales, encompassing everything from the distances between the particles in the air we breathe to the distances between stars in a galaxy, this expansion is countered by the pull of gravity. But at the scale of the Universe as a whole, expansion overcomes gravity, and, in effect, galaxies are seen to be moving away ("receding") from one another.

The Universe's expansion also causes the light we receive from faraway galaxies (and any objects inside them) to be shifted to longer wavelengths; the more distant a galaxy is from us, the redder its light will appear. In astronomical parlance, it is said to have a higher **redshift** (figure 17). This is similar to the Doppler effect. However, while the latter is due to an object's motion in respect to the observer, cosmological redshift is due to the expansion of the Universe and its effect on the distances between objects.

The luminosity distance to the supernova, derived from the width-luminosity relation, is also dependent on the expansion of the Universe. These two properties—the luminosity distance and the redshift—are connected in

Figure 17 The light from a distant supernova will be shifted in wavelength (redshifted; dashed spectrum), relative to the light from a similar, nearby supernova (solid spectrum). Its light curve will also appear fainter than that of the nearby supernova. Together, the redshift and the difference in brightness can be used to estimate the distance to the faraway supernova.

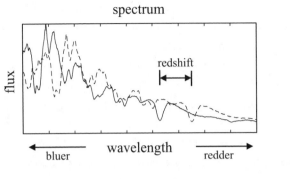

spectrum

flux

redshift

wavelength

bluer   redder

distant
(redshifted)

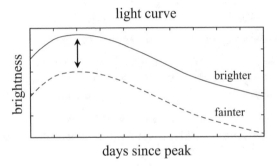

light curve

brightness

brighter

fainter

days since peak

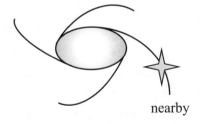

nearby

such a way that by putting the two together, we can derive the parameters of the cosmological model that describes the Universe's expansion.

In the late 1990s, two groups of astronomers used this method in a race to measure how gravity was expected to slow down the expansion of the Universe. Instead, they discovered that the rate of expansion was *speeding up*.[8] Dubbed "dark energy," whatever it is that is pushing against gravity and causing the Universe's expansion to accelerate accounts for roughly 70% of the energy density of the Universe.[9]

At first, the acceleration was attributed to the cosmological constant (denoted by the Greek letter *lambda*, Λ), a parameter that Albert Einstein had originally included in his equations of general relativity to make them produce a static Universe.[10] After Edwin Hubble showed that galaxies were receding from us, i.e., that the Universe was expanding,[11] Einstein removed the cosmological constant from his equations.

Reviving the cosmological constant is tempting but requires a physical interpretation. The most compelling suggestion so far was put forward by Yakov Zel'dovich in 1968,[12] when he argued that the very vacuum of space should have a positive energy density due to the constant creation and annihilation of particles. Although the particle-antiparticle pairs conserve energy by annihilating with each other (so that something cannot be created out of nothing), they can first interact with other particles in space and pass along some of their energy. The problem with this interpretation is that, to date, all efforts to calculate the strength of this effect are orders of magnitude off what is necessary to explain the observed acceleration.

Other explanations for dark energy include exotic formulations of gravity or reworkings of general relativity on large scales.[13] The wide range of theoretical explanations for dark energy stem from the current state of observations. As measurements of the energy density of dark energy, and whether and how it evolves with cosmic time, grow more precise and accurate, some models will be pruned away while others will be tweaked and strengthened, producing new predictions for astronomers to test.

The age of the Universe

Edwin Hubble's discovery that the Universe was expanding derived from his observation that the more distant a galaxy was from us, the faster it was seen to be receding. Hubble's law, as it is called, is simple: the recession velocity of the galaxy is equal to its distance times a constant, H_0 (pronounced H-naught), which is named after Hubble as well. Hubble's law is general, in that any observer in any galaxy in the Universe should see the same thing: that other galaxies are moving away from theirs (one day in the distant future, perhaps we will be able to test this statement). That means not only that the Universe is expanding but that there was a time in the past when all galaxies—all the matter in the Universe—were concentrated in the same point. The time between that point and the present is given by $1/H_0$ and is called the expansion age of the Universe or "Hubble time."

One of the most precise and accurate ways to measure H_0 is to compare the redshifts and distances of Type Ia supernovae in nearby galaxies.[14] The redshift of a galaxy is related to its recession velocity, so that by comparing redshifts and distances, we can once again use Hubble's law to derive H_0.[15] The latest measurement of H_0, using Type Ia supernovae discovered in nearby galaxies and anchored by variable stars in those galaxies observed with the Hubble Space Telescope, yields a value of $H_0 = 73.2 \pm 1.7$ kilometers per second per megaparsec,[16] which means that the expansion age of the Universe is $1/H_0 = 13.4 \pm 0.3$ billion years.

Dark matter and strongly lensed supernovae

Dark matter was first discovered in 1933 by Fritz Zwicky when he noticed that the galaxies in galaxy clusters were moving too fast. Without an extra, invisible source of mass to hold them in place, the galaxies would fly away in every direction.[17] Forty years later, Vera Rubin observed that stars in the outer reaches of galaxies were also moving too fast.[18] Were galaxies not embedded in halos of dark matter, those stars, like Zwicky's galaxies, should have been flung off.

We cannot see dark matter, but we can deduce its existence from its gravitational interaction with regular matter, as Zwicky and Rubin did, or through its effect on space-time itself. According to general relativity, any concentration of mass causes the fabric of space and time to bend. Any light traveling through space will also be bent, as if refracted by a lens. Moreover, the light will be amplified and distorted.

This **gravitational lensing** effect allows astronomers to observe distant galaxies too faint to see otherwise. If there is another galaxy in between the far-off galaxy and the observer, the light from the distant galaxy will be bent around the closer galaxy. If the two galaxies are exactly aligned, the light from the background galaxy will be smeared into an "Einstein ring" around the foreground galaxy. If the galaxies are slightly misaligned, the light from the background galaxy will be smeared into arcs or multiple images of the galaxy around the foreground galaxy (see, for example, the arcs around the large galaxy at the center of the cluster shown in the upper right corner of plate 2).

If, instead of a distant galaxy, the background object is one star—or its explosion—it will appear as four distinct, amplified images arrayed around the foreground galaxy in an "Einstein cross" configuration. If the supernova and lensing galaxy are misaligned, some of the paths the light from the supernova takes around the foreground galaxy will be longer than others, and each resultant image will show the supernova as it was at a different point in time; three of the images will essentially be "replays" of the supernova.

The strength of the lensing effect depends on the amount of mass concentrated in the lensing object; the more massive it is, the more it will warp its surrounding space-time. Galaxy clusters, which house dozens of galaxies embedded in gas and a dark matter halo, are the most massive objects in the Universe. If a Type Ia supernova were to go off in a galaxy somewhere behind the cluster, its light would be amplified by the mass of the cluster. Because Type Ia supernovae are standard candles, comparing the amplified brightness of the supernova to the actual brightness expected at its distance would reveal the strength of the lensing effect and allow us to derive how much mass—regular and dark—would be required to produce the observed amplification. Moreover, if the supernova is multiply imaged, measurements of the time delays between each image of the supernova could be used to independently derive Hubble's constant, H_0.[19]

In 2014, a multiply imaged, strongly lensed supernova was discovered with the Hubble Space Telescope. The host galaxy of the supernova was multiply imaged by the galaxy cluster, with delays of several years between each image

We cannot see dark matter, but we can deduce its existence from its gravitational interaction with regular matter.

of the galaxy. One of these images was further lensed by one of the member galaxies of the cluster, causing the supernova to be multiply imaged as well. In plate 7, this supernova, nicknamed SN Refsdal, appears as four yellow points of light arrayed in an Einstein cross.[20]

Although not a Type Ia supernova, SN Refsdal was still used by several groups to test their models of the dark matter distribution in the galaxy cluster. One of the galaxy images was thought to lie in the future of the galaxy image in which SN Refsdal appeared, prompting each group to predict when the supernova would reappear. A follow-up campaign monitored the galaxy cluster until SN Refsdal reappeared a year later.[21] This was another brilliant vindication of Einstein's theory of general relativity.

Creators, destroyers, and mappers of dust

The space between stars is not empty. If you were to weigh it, most of the matter found between stars would be gas (consisting mainly of hydrogen). But up to 1% of that matter, by mass, would be in the form of microscopic dust grains. Ranging in size between a few nanometers and a few microns (that is, between a billionth to a millionth smaller than us), dust grains are composed of heavy elements, such as carbon, oxygen, silicon, and iron. Theoretical simulations show that dust grains can form either in the pre-explosion winds of stars or in the supernova ejecta themselves as they expand and cool. However, the shocks created by supernova remnants are expected to destroy some fraction of that very same dust.[22]

It is still unclear whether supernovae are net creators or net destroyers of dust. What is clear is that they can be used to study the properties and distribution of preexisting dust in their host galaxies. Dust in the vicinity of a supernova, as well as dust swept up by the expanding supernova remnant, will radiate in the infrared. How much light will be emitted at each infrared wavelength depends on the optical properties of the dust grains. As the light from the supernova travels through its host galaxy, cold dust along the way will absorb some of the supernova's light, manifesting in absorption features in spectra of the supernova.[23]

Dust in the vicinity of the supernova could also create **light echoes**. The light from the supernova travels out in all directions. Some of that light makes its way directly to us and is picked up by our telescopes. Light traveling in other directions may encounter sheets of concentrated dust lying in space. A fraction of the light will be scattered by the dust in different directions, including toward Earth. The combined time taken by the light to reach the dust sheet and then travel toward us is greater than the time it took the original supernova light to reach us.

This means that if we observe a supernova long enough, there is a chance that at some point we will begin to receive some of its light again, delayed by its travel to the dust sheet. These light echoes will not be identical to the original light but bluer, since dust preferentially scatters bluer wavelengths of light. The differences in color and brightness between the original light and the light echo depend on the size of the dust grains. Moreover, over time the light from the supernova will encounter dust sheets located farther away from the explosion site. In astronomical images, the light echoes will appear to expand outward, like ripples in a pond. This effect allows astronomers to map the distribution and composition of the dust encountered by the light from the supernova.[24]

Hostless supernovae and invisible galaxies

On a dark, clear night, those of us living in the northern hemisphere can look up and see the bright stars and dark dusty band that is the Milky Way. In the southern hemisphere, stargazers will also see our Galaxy stretching across the sky, as well as two others: the Large and Small Magellanic Clouds. Each of these so-called "dwarf" galaxies contains fewer than a billion stars, and they are slowly being torn apart and consumed by the Milky Way.

Just as lightweight stars are more common than massive stars, so dwarf galaxies are more common throughout the Universe than large galaxies like ours. However, since larger galaxies contain more stars, they are more luminous and easier to spot across the vast distances of space. Alien astronomers living in a faraway galaxy might be able to make out the Milky Way, but the smaller, fainter Magellanic Clouds might remain invisible to them.

Supernovae can be used to discover these dwarf galaxies. When a supernova is discovered, it is usually straightforward to associate it with the galaxy in which it exploded by finding the closest galaxy with the same redshift. In most cases, the association will also be apparent by eye; the supernova will either be on top of its host galaxy or close by (plate 2). Rarely, there will be no galaxies close enough to the supernova to provide a plausible host. These "hostless" supernovae most probably exploded in a galaxy too faint to appear in the image. Longer observations, which provide deeper images, often reveal the faint, underlying galaxy; by illuminating their locations, hostless supernovae act as cosmic metal detectors.

The more we learn about supernovae, the more use we can make of them in other fields of physics and astronomy. However, our knowledge of the workings of these explosions is far from complete. In the next chapter, I describe some of the outstanding questions in the field, which impact not only our understanding of supernova physics but also the way we use these phenomena, especially in cosmology.

BURNING QUESTIONS

We have made great strides in our millennia-old quest to understand the nature of visiting stars, but, luckily enough for us, many questions remain unanswered. Answering these questions, the more pressing of which are described below, holds two promises. First, of course, is the promise to gain a better understanding of the supernovae themselves. But far more exciting is the allure of discovering brand-new types of supernovae, exotic physics, or a completely different way of thinking about the Universe.

Supernova progenitors

What type of star ends up exploding as each of the supernova types shown in figure 8? The "progenitor" question, as it is called, impacts all of the open questions presented in this chapter. Answering it will not only tell us which stars explode as supernovae but will also provide hints to how they explode, how we can use them, and the nature of the new types of explosions we keep discovering.

The best way to connect between each supernova and its progenitor is to directly observe the exploding star. This can be done if, by chance, the location of the supernova has already been imaged in the past. Such "pre-explosion" images can be compared with images of the same location after the supernova has faded away. The progenitor star should appear in the pre-explosion image at the exact location of the supernova but be absent from the post-explosion image.[1]

Pre-explosion images have been used to identify red supergiants as the progenitors of several Type IIP supernovae.[2] A blue supergiant called Sk –69°202 was identified as the progenitor of SN 1987A.[3] Yellow supergiants, perhaps in binary systems, were connected to several Type IIb supernovae.[4] And luminous blue variable stars (massive blue stars whose brightness fluctuates over time) were found to be the progenitors of Type IIn supernovae (figure 18).[5] A blue object at the position of the Type Iax supernova SN 2012Z is consistent with a binary system comprising a white dwarf and a helium star, but we are still waiting for the supernova to fade away before post-explosion images can reveal whether this object has disappeared.[6] No

Figure 18 Of the stars visible in a pre-explosion image from 1997 (bottom left), the star marked by a white circle, identified as the progenitor of SN 2005gl (bottom center), no longer appears in a post-explosion image from 2007 (bottom right). Credit: NASA, ESA, A. Gal-Yam (Weizmann Institute of Science), and D. Leonard (San Diego State University).

progenitors have been directly detected yet for Type Ic or Type Ia supernovae.[7]

There are many other, indirect methods to probe the nature of supernova progenitors, such as charting the locations of the supernovae in their host galaxies (younger progenitors will be in bluer, more star-forming areas) or the distribution of supernova classes between different types of galaxies (core-collapse supernovae are only seen in star-forming galaxies, while Type Ia supernovae are discovered in both star-forming and "passive" galaxies, which no longer form stars);[8] measuring the rates at which different types of supernovae occur and comparing them to various properties of their host galaxies;[9] pinpointing when the supernova exploded and how long it took to rise to maximum light, which holds information about the size of the exploding star;[10] and detecting (or not) the supernova in other wavelengths, such as radio and X-rays.[11] The hope is that, together, these independent techniques will each contribute enough information to paint a complete picture of the progenitors.

We still do not have a good handle on the progenitors of Type Ia and Type Ib/c supernovae. There is a wide consensus that the star that explodes as a Type Ia supernova is a carbon-oxygen white dwarf, but something must be done to destabilize the star and cause it to explode. There are several suggested explanations,[12] the majority of which invoke the observation that most stars exist in binary (or more complex) systems, where the white dwarf may interact with a companion star.

In one set of models, the companion star can be a hydrogen-burning star or a helium-burning red giant.[13] Due to gravitational interaction between the two, the white dwarf will siphon gas from its companion star. The white dwarf's mass will increase, causing it to shrink in on itself. This will raise the pressure and temperature in the core, eventually leading to carbon ignition.

In another popular scenario, two white dwarfs orbit each other, drifting closer together until, after dancing around each other for millions of years, they finally merge. The new, massive white dwarf is once again dense and hot enough that the carbon in its core is ignited and the explosion takes off.[14]

A common misconception is that white dwarfs will explode as Type Ia supernovae when they reach the so-called Chandrasekhar mass. This is the mass at which the electron degeneracy pressure can no longer counteract gravity, causing the white dwarf to collapse into a neutron star.[15] Yet, unlike core-collapse supernovae, Type Ia supernovae are not caused by such a collapse but by the explosive burning of carbon.[16] It just so happens that in a "vanilla" white dwarf the mass at which the temperature and pressure in the core are high enough to ignite carbon is very close to the Chandrasekhar mass. However, some explosion models allow the white dwarf to explode at masses below or above the Chandrasekhar mass (these are called, respectively, "sub-Chandra" and "super-Chandra" models).

Instead of core collapse, carbon ignition leads to a runaway thermonuclear explosion. White dwarfs lack the thermostat-like behavior of regular stars, where the nuclear reactions in the core regulate the temperature and size of the star (chapter 4). Unlike the gas pressure that counteracts the gravitational pressure in regular stars, electron degeneracy pressure does not depend on temperature, so even though the nuclear reactions triggered by the ignition of carbon in the white dwarf's core raise its temperature, the white dwarf will not regulate itself by expanding. This failure leads to a runaway process where the rising temperature leads to an enhancement of the nuclear reactions, which further raises the temperature. In a matter of seconds, the buildup of energy released by this thermonuclear runaway process leads to an explosion strong enough to completely destroy the star.

For Type Ib, Ic, and IIb supernovae, the best progenitor candidates are either single, massive stars that lose most of their outer envelopes through winds or binary systems in which one star strips the other through gravitational interaction.[17] A rare subtype of Type Ic supernovae, called Ic-BL for the broad emission features seen in its spectra, holds a further mystery; some of these events have been linked to **gamma-ray bursts**, seconds-long bursts of high-energy photons that are still poorly understood.[18]

Singles, binaries, triples

The debate surrounding the progenitors of Type Ia and Ib/c supernovae brings up an issue that has long been sidelined in most of astrophysics. To paraphrase John Donne, "No star is an island entire of itself."[19] While many stars, like our Sun, are *single*, i.e., they have no companion stars and have evolved in perfect isolation, observations show that many stars are born in binary systems and a small fraction are born in triple or higher-order systems.[20] Binarity is more common among the massive stars that die in core-collapse supernovae, and more than 70% of massive stars in binaries are expected to interact with each other.[21] So how does companionship affect a star's life?

The effects of multiplicity can profoundly change the composition, appearance, and lifespan of a star. If the stars are widely separated, they will barely affect each other. But if they are close enough to each other, the gravitational attraction between them will deform their shapes, transforming them into egg-shaped ovoids. Closer still and the denser star will begin to syphon gas off of its companion, changing both of their compositions. In extreme cases, the stars will merge (or "form a common envelope" in astrophysical jargon). The latter two cases can lengthen the lives of massive stars, delaying core-collapse explosions.[22] As I noted above, most Type Ia supernova progenitor models also assume some interaction between the white dwarf

Many stars are born in binary systems and a small fraction are born in triples or higher-order systems.

and a companion star. The physics of the phase when the two stars have merged is still being worked out.[23]

Triple systems have become popular subjects of study over the last few years, mostly because of the Lidov-Kozai mechanism,[24] in which a distant third star causes the distance between the stars in the inner binary to grow larger and smaller periodically. This makes it more likely that the stars in the inner binary will interact with one another and, in extreme cases, even collide. Lidov-Kozai oscillations have also been proposed as another channel for producing Type Ia supernovae.[25]

Failing to explode

Figuring out which stars explode as which types of supernovae is not enough; we still need to explain *how* the stars blow up. Although we have begun to piece together a coherent picture of how core-collapse and Type Ia supernovae explode, there are still several gaping holes in our understanding of the physics involved. We know this from theoretical simulations of supernova explosions, which have yet to produce explosions from start to finish.

Every explosion model faces the same challenges: produce a simulation that shows how an explosion occurs from first principles and recreate the observed supernova light curves and spectra. As with any physics problem, theorists start off with the simplest possible simulations, using well-understood and simple-to-code physics. The results of this first wave of simulations inform the design of the next set of simulations, in which more complicated physics is added (for example, magnetic fields or general relativity), initial conditions and assumptions are altered (such as the mass and composition of the simulated star), or different mathematical methods are incorporated into the code to allow for faster or more detailed simulations.

Gradually, as computation power grows ever cheaper, the simulations advance from charting the progress of the explosion in one dimension to two and finally three dimensions. Some simulations, which produce successful explosions in one or two dimensions, fail in three dimensions, hinting at either a need for the inclusion of more physical processes in the simulation or a deeper understanding of the physics already included. The products of these simulations—synthetic light curves and spectra, yields of different elements, and more—are then compared to observations, and the simulations are revised accordingly. Both revised models and brand-new ones will try to make predictions that observers can test. Successful models gain more interest and credibility, while unsuccessful ones are either revised or pruned away.

Core-collapse supernovae were initially thought to be the result of a shockwave generated by the newborn neutron star at the center of the collapsing star. This shockwave was supposed to be powerful enough to push back against the still-collapsing upper envelopes of the star and blow them off into space. However, simulations consistently show the shockwave losing energy and stalling ~150 kilometers above the center of the star.[26]

Current theoretical work focuses on finding the right physical process that could revive the shock and send it plowing through the star. The most promising candidates are the neutrinos created by the nuclear processes that turn the iron core into a neutron star (see note 8 in chapter 4). These low-mass, subatomic particles carry with them a large fraction of the energy produced by the collapse of the star; if only 0.1% of that energy were to be passed on by the shockwave to the outer envelopes of the star, that would be enough to blow them away.[27] However, though numerous, neutrinos barely interact with other particles, making it difficult to pass on their energy.

Type Ia supernova explosions have their own set of problems. In order to match the elemental signatures observed in Type Ia supernova spectra, simulations need to create both intermediate-mass elements up to silicon as well as the heavier iron-group elements. However, synthesizing the first group of elements requires a low-density environment while the second group requires a high-density environment.

A long-standing solution to this problem is to assume that the explosion of the white dwarf happens in two phases. As noted in chapter 5, the explosion starts out with a blast wave that moves outward through the star at velocities lower than the speed of sound. During this subsonic deflagration phase, the density of the white dwarf is low enough to create the intermediate-mass elements. Then, for an as-yet-unknown reason, the explosion transitions into a detonation phase during which the blast wave accelerates to velocities higher than the speed of sound (supersonic). The shock can now raise the density of the matter it passes through to values high enough to create the iron-group elements.

It is unknown how, exactly, the white dwarf transitions from deflagration to detonation. In simulations, this transition is written into the code explicitly instead of developing naturally. While some groups continue to explore ways in which the transition between the two phases could happen on its own, others are developing new explosion models.[28]

Standardizing the standard candles

Though we have yet to fully understand which stars explode as supernovae and how those explosions occur, we already use some types of supernovae as tools (chapter 6). Without clear answers to the first two questions, can we be sure that we are using them correctly?

Until a decade ago, the main source of uncertainty when using Type Ia supernovae as distance indicators was the number of explosions discovered in a given survey. As surveys have grown ever larger, this source of statistical uncertainty has diminished in importance. Now the use of Type Ia supernovae for experiments in cosmology is dominated by systematic uncertainties that have to do with the way we use these supernovae as tools and how we design our experiments. The area of most concern, though, is the question of how standard Type Ia supernovae really are.

Figure 16 presents an idealized version of Type Ia supernovae as standard candles where, after accounting for the correlation between the peak luminosities and widths of the light curves, the light curves perfectly align with each other and the scatter in brightness disappears. In reality, after accounting for this correlation, as well as an additional correlation between luminosity and color (more luminous Type Ia supernovae are bluer than less luminous ones), there still remains an *intrinsic* scatter in luminosity, which translates to an uncertainty of ~8% in the distances derived from Type Ia supernova observations.[29]

Some astronomers think this intrinsic scatter is due to dust along the line of sight to the supernovae.[30] Some galaxies are dustier than others, and dust is known to affect both brightness and color. Others think the issue is more fundamental. The discovery of the accelerating expansion of the Universe, described in the previous chapter, rested on the assumption that the Type Ia supernovae we see in nearby galaxies are identical to the ones we see in faraway galaxies. But, as long as the nature of the progenitor system of Type Ia supernovae remains unclear, this assumption is open to challenges.

Because the speed of light is finite, light from distant objects takes millions of years to reach us. So, when we look at a distant galaxy, we see it as it was millions or billions of years ago. The stars that reside in nearby galaxies, seeded with heavy elements created by successive generations of supernovae, have a different chemical composition from what we see in stars in distant galaxies, which show the conditions of an earlier period of the Universe's evolution. It is quite possible that the progenitors of Type Ia supernovae also evolve over time, and that the explosions we see in nearby galaxies and those we see in distant ones are intrinsically different in ways we still do not fully comprehend.

Further correlations can be used to reduce the intrinsic scatter in luminosity. For example, brighter Type Ia supernovae have been found to preferentially explode in larger, more massive galaxies while fainter events are more often found in smaller, less-massive galaxies.[31] The problem with this approach is that it is purely empirical, i.e., it is based solely on observations and not on any deep physical understanding of the explosions. It leaves a nagging feeling that, like a chainsaw that we borrowed from a neighbor, we are using a tool that we do not fully understand or control.

Exotic supernovae: predictions

In chapter 4, I told a story about the creation, evolution, and death of certain types of stars. How can we test this picture of stellar evolution? Every scientific theory has to provide predictions for observers to test. In this case, we can search for the predicted, exotic explosions of stars in the mass range of 6–10 Solar masses and of those with masses larger than 40 Solar masses.

Depending on their chemical composition, some stars in the range of 6–8 Solar masses will go through carbon burning and end up as oxygen-neon white dwarfs. Most stars in the mass range of 8–10 Solar masses will progress to neon burning and the next fusion phases, finally leading to the collapse of an iron core. However, a thin sliver of stars in this mass range may go through an **electron-capture supernova**, where the absorption of electrons by magnesium and neon atoms causes a drop in pressure support and the collapse of the star's core. The ensuing explosion is expected to look like a low-luminosity version of a Type IIP supernova and leave behind either a neutron star or an oxygen-neon-iron white dwarf.[32] SN 2016bkv is claimed to have been such a supernova,[33] but theoreticians are still split as to whether this type of explosion should occur at all.[34]

According to the simulations I cited in chapter 4,[35] in stars with masses in the range of 40–100 Solar masses, instead of forming a short-lived neutron star, the iron core will collapse directly into a black hole. Without the neutron star phase there will be no outward shock and so no supernova; these stars are expected to simply wink out of existence. Searches for such "failed supernovae" are ongoing; so far, only one candidate has been found, and the jury is still out on it.[36]

Stars more massive than 100 Solar masses are expected to undergo pulsational pair instabilities. As before, after the helium-burning phase, the star's core begins to contract. However, instead of raising the core's temperature and providing more gas pressure support against further collapse, the energy from the contraction is diverted into the production of pairs of electrons and positrons (the electron's antiparticle, identical in every way except for a positive electric charge), further accelerating the collapse. In stars that were initially 100–140 Solar masses, this mechanism is thought to cause periodic violent pulses that eject the outer envelopes of the star but still allow an iron core to form and collapse into a black hole. The resultant explosion is called a "pulsational" **pair-instability supernova**.

In stars with initial masses in the range of 140–260 Solar masses, a single pulse is enough to completely disrupt the star, leading to a pair-instability supernova that leaves no remnant behind. The higher the initial mass of the star, the brighter these supernovae are predicted to be, with the brightest explosions outshining the most luminous supernovae known today.

Above 260 Solar masses, the star is once again expected to collapse directly into a black hole without producing a visible supernova. Whether or not very massive stars will explode as pair-instability supernovae depends on the initial chemical composition of the star. The higher the fraction of heavy elements in its composition, the more mass it will lose throughout its lifetime to winds. If it loses too much mass, it will not explode as a pair-instability supernova.

A few superluminous supernovae (see below) have been identified as potential pair-instability explosions of stars more massive than 100 Solar masses, but these identifications are still debated.[37] However, the dependence of pair-instability supernovae on the chemical composition of their progenitor stars has led theorists to assume that most, if not all, pair-instability supernovae will originate from the first generation of stars. Only these stars, called "Population III" stars, would be chemically clean enough to produce very massive stars with winds weak enough not to remove too much mass prior to explosion.

The scarcity of heavy elements in the early Universe holds out the possibility that some Population III stars could come in at more than 10,000 Solar masses. The evolution of these supermassive stars is highly uncertain; they might collapse into massive black holes, lose most of their mass to strong winds, or explode as pair-instability supernovae. These exotic stars—and their even more exotic explosions—are too far away in the Universe's past to be observed with the current generation of telescopes, but they just might be visible to the next generation of extremely large telescopes described in the next chapter.

Exotic supernovae: observations

"There are more things in heaven and earth, Horatio," said Hamlet, "than are dreamt of in your philosophy."[38] Hamlet, of course, was an observer. While observers search for the pair-instability and electron-capture supernovae predicted by theorists, those same theorists have to account for exotic supernovae that do not fit into any of their existing models. Some of these may necessitate the development of brand-new physics.

Superluminous supernovae are the brightest supernovae observed today, with luminosities 10 to 100 times higher than those of normal core-collapse supernovae. As with regular supernovae, some superluminous supernovae exhibit hydrogen in their spectra and are dubbed Type II, while those devoid of hydrogen are called Type I. Some of the events in this class can be explained as explosions of stars more massive than 100 Solar masses, which produce large amounts of radioactive nickel and hence higher luminosities. Some Type II superluminous supernovae have narrow hydrogen features in their spectra, indicative of interaction with gas previously expelled by the star. However, the energy produced by such interaction is not enough to explain these events' abnormally high luminosity.

A more exotic source of power is the neutron star or black hole created by the core collapse. Through different physical processes, these so-called "central engines" could continually inject energy into the expanding supernova ejecta and raise the explosion's luminosity.

Since most stars start out rotating, and have a magnetic field, the neutron stars they create will also rotate and be magnetized. If, in rare cases, the rotation is fast enough and the magnetic field strong enough, the neutron star will be known as a **magnetar**. Over time, the rotation of a neutron star slows down. As this happens, the neutron star releases its rotational energy in a wind of high-energy photons and pairs of electrons and positrons that deposit their energy into the ejecta. Once the magnetar has spun down, no more energy will be injected into the ejecta and the supernova will fade away.

In more massive stars, the central engine may be a black hole. Though most of the star will explode outward, some of its inner material may not be energetic enough to escape the pull of the black hole and will fall in. As with the magnetar, because the star had some rotation before it exploded, the infalling matter will orbit around the black hole in a disk of swirling hot gas before eventually falling in. The disk will be hot but will not be able to cool down efficiently, instead launching either a wind or a jet, which once again will interact with the expanding ejecta and inject the necessary energy to power the light curve.[39]

Rapidly evolving transients is a catchall phrase for any type of supernova that quickly rises to peak (<10 days) and then quickly fades away again. This is a heterogeneous group of **transients** with a wide range of luminosities; some are much dimmer than regular supernovae while some are more luminous. Unlike regular supernovae, several of the events in this class have light curves devoid of a radioactive tail like the ones shown in figure 9. Each of the solutions proposed for superluminous supernovae—radioactive decay, interaction with shells of nearby gas, or magnetars—can explain some but not all of the supernovae in this class. It is also possible that several physical processes are at work simultaneously or that a completely new power source is required to explain these rare events.

Popular accounts of science often strive to pinpoint one question, one puzzle piece that, once slotted in, would finally solve the puzzle. This is a simplistic approach that does not do justice to the complexity of the Universe we study or to the hard work and ingenuity of the people engaged in those studies. It is true that all of the questions I described in this chapter, like most of astrophysics, are interconnected. Answering one would affect how, and how quickly, we answer the others. But no single question holds the key to answering all the others. And that is a good thing; it means we get to keep asking questions (as well as our jobs). As we will see in the next chapter, the next decade promises to be an exciting one for supernova research, and some of the questions posed here may be answered sooner rather than later.

A BRIGHT FUTURE

The future of supernova studies is bright (pun intended). The coming decade will see the construction of the biggest telescopes ever built, the launch of successors to the Hubble and other space telescopes, and the beginning of the Rubin Observatory Legacy Survey of Space and Time. Using these together with the now fully operational gravitational-wave observatories, we will boldly stride into a new era of multimessenger, big-data, **time-domain astronomy**. In this final chapter, I describe how these new observatories and buzz words will transform the study of supernovae and how you too can join the revolution.

Expanding the observational phase space

In 1975, Martin Harwit published a philosophically minded paper in which he argued that the number of distinct astrophysical phenomena in the Universe was finite and could be estimated.[1] Though an intriguing idea, it is now mostly forgotten. Yet this paper continues to be a bedrock of modern astrophysics because of a different statement made by Harwit en route to his conclusion: expand our observational phase space and new discoveries will follow.

In physics, "phase space" is the area described by any two variables. For example, the phase space of length and weight of cats would be a two-dimensional plot in which weight was one axis and length the other. Some cats would be short and fat, others long and skinny. All the possible combinations would be represented as points on this plot and would be said to inhabit the "feline weight–length phase space."

Harwit plotted three phase space diagrams that showcased the capabilities of the observatories then in operation. The x-axis for all of these plots was the wavelength of light received by the telescopes. The y-axes were angular resolution (how easy it is to tell two objects apart when they are close together), spectral resolution (how easy it is to tell two emission or absorption features apart in an object's spectrum), and time resolution (how fast observations can be taken and recorded). The plots made it clear that large portions of these phase spaces were unreachable with the observatories of the 1970s, and served as an

Expand our observational phase space and new discoveries will follow.

impetus for the planning and construction of new observatories to fill those gaps.

NASA's Great Observatories was one program directly influenced by Harwit's paper.[2] The Hubble Space Telescope (optical and near-infrared; named after Edwin Hubble, famous for showing that distant galaxies were moving away from us), Spitzer Space Telescope (infrared; named after Lyman Spitzer, who envisioned launching telescopes into space), Compton Gamma Ray Observatory (gamma rays; named after Arthur Compton, who studied the absorption and emission of gamma rays), and Chandra X-ray Observatory (X-rays; named after Subrahmanyan Chandrasekhar, who worked out the maximum masses at which white dwarfs and neutron stars would collapse in on themselves) were all designed to open up new areas of observational phase space.

The move to space was motivated by astronomers' complicated love-hate relationship with the Earth's atmosphere. On the one hand, without the atmosphere acting as a shield against ultraviolet light, X-rays, and gamma rays, Earth would probably be a barren, desolate rock. On the other hand, it means that a large fraction of the light zooming through the Universe never reaches our telescopes. Moreover, light that does manage to filter through the atmosphere (optical, radio, and some infrared) gets smeared so that instead of sharp points of light, stars and other objects appear in our images as blurry blobs.

Situating our telescopes atop high, dry mountains reduces some of this effect, but never eliminates it completely.[3] By launching telescopes into space, we gain access to wavelengths of light unreachable from the ground and produce sharper images to boot.

With these new observatories in orbit, new discoveries quickly started flowing in, and it has become an article of faith of modern science that opening up new areas of observational phase space will, by default, lead to previously unimagined discoveries. This principle has guided the astronomical community for the last few decades and has shaped the goals and design of the next generation of observatories currently under construction.

Since the launch of the Great Observatories, astronomy has been split between ground-based and space-based telescopes. On the ground, the last century has seen a steady increase in the size of telescopes. Walk into a dark room and your pupils will expand, letting in more light and allowing you to make out your surroundings. The same is true of telescopes; larger mirrors collect more light, making it possible to see stars and galaxies too faint—whether because they are intrinsically less luminous or because they are farther away—for telescopes with smaller mirrors.

The size of a telescope's mirror also affects its resolving power, so that bigger telescopes can view the same objects at a higher resolution. Where a telescope with a small mirror will see a star cluster as a smoothed-out blob, a telescope with a larger mirror might make out individual stars. 4-meter telescopes (telescopes with mirrors with diameters of 4 meters) were first constructed in the 1970s, followed by 6-, 8-, and 10-meter telescopes in the 1990s and early 2000s.

Figure 19 shows a phase space diagram for supernovae and other transient astrophysical phenomena. The y-axis marks the peak absolute brightness of the event (more luminous objects have more negative values than dimmer ones) while the x-axis notes the time it takes an event to fade by two magnitudes from peak brightness. Due to the sensitivity of current observatories, all of the supernova types described in the previous chapters populate the right side of this phase space. To populate the left side, we need the next generation of observatories to do three things: detect fainter transients, detect transients that radiate light in wavelength ranges previously inaccessible, and detect transients that change on rapid timescales (days instead of weeks).

Bigger: Extremely Large Telescopes

The next generation of telescopes, the so-called "extremely large telescopes," or ELTs, will have mirrors with light collection areas ten times as large as those of the biggest telescopes currently in operation. Three ELTs are currently

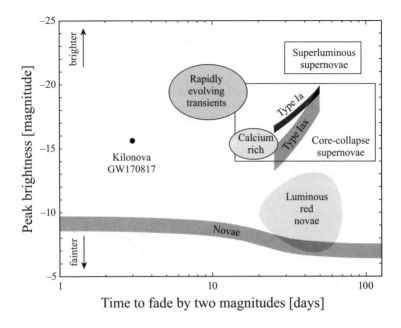

Figure 19 Supernova discovery phase space, based on the peak brightness of an event and the time it takes to fade by two magnitudes from peak. All of the supernova types known today occupy the right side of the plot. LSST and the extremely large telescopes will populate the left side. Source: adapted from S. A. Rodney et al., "Two Peculiar Fast Transients in a Strongly Lensed Host Galaxy," *Nature Astronomy* 2 (April 2018): 324–333.

under construction: the 24.5-meter Giant Magellan Telescope at Las Campanas, Chile; the Thirty Meter Telescope on Mauna Kea, Hawaii; and the 39.3-meter European Extremely Large Telescope at Cerro Armazones, Chile.

The ELTs will be able to observe farther back in time than ever before, but they will still be too small to detect the light emitted by the very first stars. They will, however, be able to detect the light from the exotic explosions of those stars (chapter 7).

In space, the Hubble Space Telescope, which has been in operation for more than 25 years, is set to be replaced by a new observatory, the James Webb Space Telescope (named after NASA's director during the 1960s, who maintained a balance between human space flight—the Apollo program—and the launch of unmanned spacecraft to explore the Solar System). Like the ELTs, the James Webb will have a larger mirror (6.5-meter) than Hubble's (2.4-meter), providing it with a larger light collection area and higher resolving power.

Faster: Rubin Observatory Legacy Survey of Space and Time (LSST)

Starting in 2022, Vera C. Rubin Observatory (figure 20), named after a trailblazer for women in astronomy and a codiscoverer of dark matter (chapter 6), will embark on

Figure 20 Rendering of Vera C. Rubin Observatory. Credit: LSST Project/ NSF/AURA.

the ten-year-long Rubin Observatory Legacy Survey of Space and Time (LSST)—the largest astronomical survey ever attempted.

Rubin Observatory will not operate as a standard observatory to which astronomers apply for time to observe their own specific targets. Instead, every night LSST will visit a thousand different parts of the southern hemisphere's night sky. In each visit, in a given filter, it will take two fifteen-second-long exposures. Even though the exposures are short, Rubin Observatory's large, 8.4-meter mirror will ensure that the images produced by LSST will be deeper than those taken by previous surveys, such as the industry benchmark Sloan Digital Sky Survey (SDSS), which uses a smaller, 2.4-meter telescope.[4] A single LSST exposure will be ten times deeper than SDSS; the ten-year stacked LSST image will be more than a hundred times deeper.[5]

LSST presents a new challenge for astronomers: we will soon be inundated by the data it produces. While most astronomers still take their own images and spectra, a growing number use large data repositories such as SDSS to either supplement their own data or replace them completely. LSST is expected to produce 20 terabytes of data each night (at the time of writing, a standard laptop had a hard drive of 0.5–1 terabytes). The images will be processed immediately and scanned for new objects or changes in existing ones. Once a minute, an alert will be sent out to servers around the world with a summary of these discoveries. Today's supernova surveys discover hundreds of supernovae and other transients per year; LSST will discover 10,000 new transients *every single night*.

LSST will affect every astronomer studying supernovae and other transient phenomena. Supernova population studies are currently based on hundreds of supernovae, at best; with LSST, similar studies will include tens of

Today's supernova surveys discover hundreds of supernovae and other transients per year; LSST will discover 10,000 new transients every single night.

thousands of objects. From studying supernovae on time scales of days and years, we will transition to studying how they change minute by minute. Rare and exotic supernovae will become commonplace. And the very ways in which we study supernovae will change.

For several years now, a large fraction of the astronomical community has been working out how to digest LSST's unprecedented data stream (often likening it to drinking from a gushing fire hydrant). The current generation of supernova surveys is already partially automated. "Pipelines"—computer programs that run continuously—process the data from the survey telescopes, produce difference images, and detect supernova candidates. The candidates are vetted by astronomers, who also decide which candidates to follow up with more imaging or spectroscopy with other telescopes.

In the LSST era, these final, human steps will have to be computerized and automated as well. First, there will be too much data for astronomers to sift through by eye. Second, in order to study rapidly evolving transients and populate the left-hand side of figure 19, follow-up decisions will have to be made quickly. And these decisions will be expensive; there are simply not enough telescopes in the world to follow up each new candidate detected by LSST.

To solve these issues, several teams are hard at work coding "brokers," algorithms designed to receive data from

LSST, process and analyze them in real time, prioritize targets for follow-up, and trigger follow-up observations with robotic telescopes.[6] Although we aim to make this process as automated and fast as possible, with artificial intelligence making many decisions along the way, human astronomers will still be indispensable. It will be up to us to decide which classes of transients to prioritize for follow-up and what kinds of queries to run on the data. Artificial intelligence grows more powerful every year and can be used to answer many scientific questions. But *asking* those questions—at least for now—remains a human trait.

Different: beyond visible light

Most supernovae are first discovered in the visible part of the electromagnetic spectrum where most telescopes operate, and are then followed up with observations in other wavelengths. In recent years, several collaborations have flipped this formula.

In 2017, the Karl G. Jansky Very Large Array Sky Survey (named after Karl Guthe Jansky, who discovered that the Milky Way Galaxy was emitting radio waves) started searching for transients in the radio region of the electromagnetic spectrum.[7] There is reason to hope for new discoveries in this wavelength range, as this is where **fast**

radio bursts, a mysterious new type of transient, were discovered in 2007.[8] Supernova remnants have long been observed at radio wavelengths. Since 2013, they are also being scrutinized in the millimeter and submillimeter regimes using the Atacama Large Millimeter/submillimeter Array (ALMA). At these wavelengths, longer than traditional radio, ALMA can trace the distribution of cold dust created directly by the supernova.[9]

Supernovae whose visible light is obscured by dust can pop out in infrared light. Hubble's successor, the James Webb Space Telescope, will extend Hubble's capabilities to longer wavelengths; where Hubble can see into the near-infrared, James Webb's cameras will see into the mid-infrared. And in 2025, NASA will launch a second infrared telescope, named after Nancy Grace Roman, the agency's first chief astronomer and "mother" of the Hubble Space Telescope. One of the Roman Space Telescope's science goals is a five-year supernova survey in visible and near-infrared wavelengths, designed to discover highly redshifted Type Ia supernovae. The Universe's expansion means that the redshifted supernovae we discover in infrared light will be more distant than those we discover in visible light. Because of the finite speed of light, light from distant objects takes longer to reach us. With these supernovae, whose light has been traveling towards us for billions of years, we will probe the Universe's expansion when it was young and gravity still held dark energy at bay.

Beyond light: gravitational waves and multimessenger astronomy

Light is not the only way for us to comprehend the Universe. On 14 September 2015, the Laser Interferometer Gravitational-Wave Observatory (LIGO) directly detected gravitational waves emitted by the merger of two black holes.[10] A prediction of Albert Einstein's general relativity since 1918,[11] gravitational waves are caused when dense, massive objects such as black holes or neutron stars accelerate through space-time. As gravitational waves propagate, they stretch and squeeze space itself.

The LIGO detectors are built specifically to detect this effect. Each detector is composed of two tunnels connected at a 90-degree angle, forming an L-shape. A beam of light is simultaneously sent down each tunnel, at the end of which is a mirror suspended from a pendulum. The light beams are reflected back down the tunnels and combined back at the source. If the tunnels are the same length, the combined light will produce a specific pattern. If, however, a gravitational wave were to pass through the detector, the length of the tunnels would change as space itself was stretched and squeezed. Each light beam would travel a slightly different distance and, when the beams were combined back at the source, they would produce a different pattern. The reason that it took nearly a century to directly detect gravitational waves is that their effect is minuscule. For LIGO to discover that first wave in 2015, it had to be able to detect a change in the length of the tunnels of just 10^{-19} meters, ten trillion times smaller than the motes of dust hovering in the air around you.

The gravitational waves detected by LIGO represent a new "messenger" of information for astronomers. It is often unappreciated that astronomers use more than just light to study the Universe. Granted, most of our telescopes are designed to gather light at various wavelengths, but we have other means of gathering information about the Universe.

Charged particles, such as cosmic rays accelerated by supernovae, can be studied using specialized detectors or through their impact with the Earth's atmosphere, which produces showers of light (called Cherenkov radiation) that can be picked up by dedicated telescope arrays.

Neutrinos, created by radioactive processes in stars and supernovae, are subatomic particles that zip through the Earth, barely interacting with anything. Special observatories, filled with water, are now able to routinely detect the light produced when a tiny fraction of these neutrinos happen to interact with the water molecules. A total of 25 neutrinos from the nearby supernova SN 1987A were detected a few hours before the first light from that supernova was picked up by regular telescopes.[12] The next time a supernova explodes in our Galaxy, it will first be observed— and perhaps even discovered—by neutrino observatories.

The true promise of **multimessenger astronomy** lies in the combination of information from the various messengers at our disposal. The neutrinos created by supernovae, for example, directly probe the radioactive processes that take place during the explosion. Likewise, some types of supernovae are predicted to produce gravitational waves.

Core-collapse supernovae are expected to produce gravitational waves at two key stages of the explosion: the rapid rotation of the iron core before it collapses and the bounce of the core as it hardens into a neutron star.[13] Type Ia supernovae are also expected to produce gravitational waves if their progenitors are merging white dwarfs. Such binary white dwarf mergers are expected to be the main source of background signal for the Laser Interferometer Space Antenna (LISA), a next-generation, space-based gravitational-wave observatory planned by the European Space Agency.

The power of multimessenger astronomy was made clear with the discovery of GW170817, a gravitational wave signal from the merger of two neutron stars (plate 8).[14] Such a merger was predicted to produce not only gravitational waves but also a **kilonova** (or "macronova"), a type of faint, short-lived transient that had never been seen before (compare its location in figure 19 to that of normal supernovae). An unprecedented follow-up campaign by seventy observatories around the world managed to detect the kilonova associated with the GW170817 gravitational-wave signal across the electromagnetic spectrum: radio, infrared, optical, ultraviolet, and X-rays (perhaps even gamma rays). These observations bore out many of the predictions about kilonovae, including their being a major source of heavy elements, from gold and platinum to uranium and plutonium.[15]

In chapter 6, we saw how Type Ia supernovae can be used as standard candles to measure distances to faraway galaxies and probe the age of the Universe and how fast it is expanding. Gravitational-wave signals can similarly be used as "standard sirens" ("sirens" because gravitational-wave signals are not a form of light; instead, they have been likened to audio signals). As with Type Ia supernovae, the age of the Universe is derived by comparing the distance and redshift of the event. In the standard siren technique, the redshift is also determined from the galaxy in which the event took place, while the distance is derived from the amplitude of the gravitational wave signal.[16]

Amateur astronomers and citizen science

You too can discover and observe supernovae. Although most supernovae are discovered by professional astronomers in dedicated surveys, many bright, nearby supernovae are routinely discovered by amateur astronomers.

In this case, "amateur" only means that the astronomer in question is not affiliated with an academic institution, in the same sense that Sherlock Holmes, who was not a police inspector, was an amateur detective. Most amateur astronomers use their personal telescopes to either scan the skies or image specific, well-known nearby galaxies. By comparing new images with old ones, and in some cases relying on their knowledge and memories of those galaxies, they often discover new supernovae at the same time as—or even before—the professionals.

Once a new supernova candidate is discovered, it is reported to the worldwide supernova community by submitting a report to the International Astronomical Union's Transient Name Server.[17] The candidate will receive an AT (Astronomical Transient) designation. If, for example, it is the first supernova candidate reported in 2022, it will be named AT 2022A. Once all 26 letters of the alphabet have been used, the next candidate will be given lowercase double letters, e.g., the 27th candidate will be named AT 2022aa, the 28th AT 2020ab, and so on. Supernova surveys are now so prolific that designations stretch to three letters; e.g., AT 2020nlb is the 9,778th supernova candidate discovered in 2020. Once a spectrum confirms that the candidate is a supernova, its designation will change from AT to SN (supernova). If it was named AT 2022cg, it will now be named SN 2022cg.

You too can discover and observe supernovae.

New supernova candidates are also reported through the publication of an Astronomer's Telegram, or ATel.[18] These short missives report the location and brightness of the supernova candidate so that other astronomers can train their own telescopes at the object and collect more data on it. Follow-up ATels often include information about the spectroscopic classification of the candidate or observations in other wavelengths. The Latest Supernovae website curates a list of all active supernovae brighter than 17th magnitude and so observable with small telescopes.[19]

Anyone can search the Transient Name Server and sign up for the Astronomer's Telegram mailing list, and amateur astronomers are welcome to register as well. Latest Supernovae is also a free, public resource, and is always looking to feature images of supernovae taken by both professionals and amateurs.

The emergence of citizen science has also made it possible for amateurs to take part in professional astrophysics experiments. Zooniverse,[20] founded by astronomers looking to classify galaxies in images taken by the SDSS in a project called GalaxyZoo, has grown into a portal that features dozens of professional science projects that you can join, from astronomy through ornithology to weather studies and history. Many of the projects on Zooniverse have resulted in peer-reviewed science publications that would not have been possible without the help of the thousands of volunteers who log on to the website. Some projects have included searches for supernovae in images taken by professional surveys. The latest, Supernova Hunters, was launched in 2016 to search for supernovae in the Pan-STARRS1 survey.[21]

You can also opt to let your computer do the work for you. Developed by the University of California, Berkeley, the Berkeley Open Infrastructure for Network Computing (BOINC) is a free piece of software that allows your computer to process data from scientific projects that require large amounts of computing time.[22] Instead of running on supercomputers, these programs are broken into chunks that are then processed by thousands of volunteer computers around the world. You can choose to let your computer receive and process data from projects in a specific science field (such as astronomy), or actively choose projects to which you would like to contribute computing time. Perhaps the best-known of these is SETI@Home, which analyzed radio observations for signs of extraterrestrial intelligence.

Finally, if you are still in school or college, there may be internship opportunities at the astrophysics department of your local university or college. Many US astronomy departments take part in the National Science Foundation's Research Experience for Undergraduates (REU) program,[23] which provides summer internships at research institutions across the US. High school students should look for science research mentoring programs,[24] which pair students with scientist mentors for summer or yearlong internships. Three good resources for the latter type of program are the Global SPHERE Network, Makers+Mentors Network, and the New York Academy of Science's Global STEM Alliance.[25]

CONCLUSIONS

In this book, I have tried to succinctly describe the essentials of our current understanding of supernovae. We have learned a great deal since Chinese court astrologers first started noting the appearance of "guest stars" two thousand years ago. Slowly but surely, we are piecing together the puzzle of how stars explode and how those explosions affect other astronomical phenomena. Our understanding of supernovae has allowed us to use them as tools in our studies of the Universe and has revealed how we owe them our very existence.

There is still a long way to go to complete the supernova jigsaw; some of the pieces do not quite fit together while others are missing altogether. Crucially, this puzzle is not being worked on by one or two gifted individuals, as science is usually portrayed. Hundreds of astrophysicists spread around the world are all working on different pieces of the puzzle, together, like a family sitting around the dining room table on a cold winter's night. There will be arguments and voices will be raised, but there will also be hot cocoa (and lots and lots of coffee), and eventually we will solve the puzzle together.

Astrophysics is experiencing a golden age. Over just a few decades, we have discovered the existence of dark matter, dark energy (using supernovae), exoplanets, and gravitational waves. We build ever-larger telescopes capable of peering farther into the Universe and deeper into its history. Some of our telescopes orbit the Earth, while unmanned spacecraft travel to other planets and beyond the Solar System.

The next decade will see a sea change in the study of supernovae and other transient astrophysical phenomena. Our eyes, augmented by the mirrors of our telescopes, are no longer the only way to comprehend the Universe. The era of multimessenger astrophysics, when we combine information from light, neutrinos, cosmic rays, and gravitational waves, is already upon us. LSST will further transform astrophysics by bringing time-domain astronomy—the study of how the Universe changes from day to day—to the fore. With ten thousand new transients discovered every night, LSST will transform our understanding of supernovae and the very methods we use to study them. I sincerely hope that in ten years, twenty at most, this book will be woefully out of date.

It is an exciting time to be an astrophysicist, whether professional or amateur. For the first time in the history of science, many of our experiments are open to participation by members of the public. You too can study supernovae and make new discoveries, whether as part of a citizen science project or by setting up your own telescope and survey at home. Finally, remember the astronomers

Astrophysics is experiencing a golden age.

of antiquity, whose knowledge of the night sky was so extensive that they could look up and detect the presence of a new star with nothing by their eyes. The next supernova in our own Milky Way Galaxy is long overdue. So, every once in a while, look up. With a little bit of luck, you could be the first person to witness the next Galactic supernova.

ACKNOWLEDGMENTS

This book is the result of the inaugural 2018 MIT Press Pitchfest in Boston. I am grateful to Jermey Matthews, Haley Biermann, and Matthew Abbate for shepherding this book to publication. I thank Noga Ganany, Mina and Dan Graur, Carles Badenes, Daniel Eisenstein, Robert Kirshner, Dan Maoz, Steven Rodney, Silvia Toonen, Dror Weil, and the anonymous reviewers for discussions and comments. While writing this book, I was a National Science Foundation Astronomy and Astrophysics Postdoctoral Fellow, under grant AST-1602595, at the Center for Astrophysics | Harvard & Smithsonian; a Research Associate at the American Museum of Natural History; and a Senior Lecturer in Astrophysics at the Institute of Cosmology and Gravitation at the University of Portsmouth. I was also hosted by the Institute of Astronomy at Cambridge University as a visiting scholar.

UNITS AND
ELEMENTARY PARTICLES

Units
- -

All units are given to a precision of two significant digits.

Gram (g)
A unit of mass. A kilogram (kg) is equal to 1,000 g. In the imperial system of units, one pound equals 454 g.

Solar mass ($M_\odot$)
The mass of the Sun, equal to 2×10^{33} g.

Centimeter (cm)
A unit of length, equal to 0.4 inches. 1 meter (m) is equal to 100 cm or 3.3 feet.

Astronomical Unit (AU)
A unit of length, defined by the distance between the Earth and the Sun. Equal to 1.5×10^{13} cm.

Light-year (lyr)
A unit of length, defined as the distance traveled by light in one year. Equal to 9.5×10^{17} cm.

Parsec (pc)
A unit of length, equal to 3.1×10^{18} cm or 3.3 lyr. Distances in the Milky Way Galaxy are measured in thousands of pc (kiloparsecs; kpc). Distances to other galaxies are measured in millions of pc (megaparsecs; Mpc).

Magnitude (mag)
A unit of brightness. Brighter objects have smaller magnitude values than fainter objects, and negative values are allowed.

Elementary particles mentioned in the text

Photon (γ)   The particle of light. Can be emitted or absorbed by atoms and has no mass.

Proton (p)   A particle with a positive electric charge. A chemical element is defined by the number of protons in the nucleus of its atom.

Neutron (n)   A particle with no electric charge that, together with the proton, makes up the nucleus of an atom. Different isotopes of a chemical element will have the same number of protons but a different number of neutrons.

Electron (e^-)   A particle with a negative electric charge. In an atom, electrons orbit the nucleus.

Positron (e^+)   The antiparticle of the electron, identical but for its positive electric charge.

Neutrino (ν)   A particle with no electric charge and a very small, as yet unmeasured, mass. It is emitted by atoms during many radioactive processes.

Baryonic matter
The "normal" matter that makes up everything we see and interact with in the everyday world. Only ~5% of the total mass-energy density of the Universe is composed of baryonic matter. Also see **Dark matter**.

Big Bang
The beginning of time and the moment when the Universe began to expand, according to the currently favored cosmological model. At that moment in time, all of the matter, energy, and space in the Universe were condensed into one point. When the Universe began to expand, it did not expand into anything; instead, the space that makes up the Universe began to stretch and the distance between any two points grew larger.

Black hole
An object that, due to the force of gravity, has collapsed in on itself. Black holes are so dense that nothing, not even light, can escape them. Small black holes, with masses similar to those of stars, are created by certain types of supernovae. Supermassive black holes, with masses millions to billions of times the mass of the Sun, are thought to reside in the centers of all galaxies.

Brightness
The amount of light received from an astrophysical object. Also see **Luminosity**.

Brown dwarf
Objects with less mass than 1% of the mass of our Sun, but more massive than Jupiter, the largest planet in the Solar System. As in stars, the gravitational collapse of brown dwarfs is halted by the pressure of the dense, hot hydrogen in their cores. Unlike in stars, that hydrogen is never ignited. As brown dwarfs cool down, they shine mainly in the infrared.

Core-collapse supernova
Any type of supernova caused by the collapse of the core of a star more massive than 8 Solar masses due to loss of pressure support against the pull of gravity. Core-collapse supernovae include all Type II and Type Ib/c ("stripped-envelope") supernovae.

Cosmic rays
Highly energetic charged particles traveling at velocities close to the speed of light. Cosmic rays are thought to be accelerated by supernova remnants.

Dark energy
A still-mysterious phenomenon that counteracts gravity at cosmic scales. Because of dark energy, instead of decelerating under the pressure of gravity, the expansion of the Universe is currently accelerating.

Dark matter
An as-yet-undiscovered form of matter that has mass and interacts with normal (baryonic) matter via gravity, but does not produce or absorb light. Makes up roughly a quarter of all the mass-energy in the Universe. Also see **Baryonic matter**.

Degeneracy pressure
Pressure produced by either electrons or neutrons as they are squeezed together at high densities. Degeneracy pressure keeps white dwarfs (electron degeneracy pressure) and neutron stars (neutron degeneracy pressure) from collapsing further under the pressure of gravity.

Ejecta
The blown-up matter of a star exploding as a supernova, expanding outward at velocities of thousands to tens of thousands of kilometers per second.

Electron-capture supernova
A theoretically predicted type of supernova in which magnesium and neon atoms in the cores of stars whose mass is 8 to 10 times that of our Sun capture electrons, thus weakening the electron degeneracy pressure (see **Degeneracy pressure**) and triggering core collapse.

Fast radio burst
A still mysterious transient phenomenon of extragalactic origin observed as a radio signal that lasts up to a few milliseconds.

Galactic wind
A wave of pressure built up by expanding supernova remnants and the winds from massive stars. Galactic winds, or "outflows," can drive gas and dust out of the disks of star-forming galaxies, removing some of the fuel required for star formation.

Gamma-ray burst

A transient phenomenon that manifests in a burst of high-energy gamma rays. Short gamma-ray bursts last mere milliseconds while long gamma-ray bursts last for more than two seconds. Some Type Ic-BL supernovae have been associated with long gamma-ray bursts. The physical mechanism behind this phenomenon is still unclear.

Gravitational lensing

A general relativity effect by which light is bent as it travels through a region of space-time that is warped by the presence of a massive object, such as a galaxy or galaxy cluster. "Strong" lensing can bend the light of distant objects into arcs or "Einstein rings" around the lensing object in the foreground. It can also create multiple warped images of the background object, each of which appears at a different time. Not as dramatic, "weak" lensing causes minute shears in the appearance of galaxies that lie behind the foreground lens.

Kilonova

A brief, faint electromagnetic flare caused by the merger of two neutron stars. Kilonovae (or "macronovae") are thought to be the main source of r-process elements (created when an atom rapidly absorbs neutrons and then radioactively decays into various stable elements), including such heavy elements as gold and platinum.

Light curve

The brightness of an astrophysical object or phenomenon, measured over time. The light curves of supernovae are measured by observing them at different wavelengths, from the radio to X-rays, and are generally seen to brighten, reach peak, and then fade. The shape of the light curve, in various filters, holds information about the nature of the exploding star and about the physics of the explosion process. The characteristic shapes of light curves are also used to arrange supernovae into the different supernova classes.

Light echo

A rerun of the light from an astrophysical event, such as a supernova, created when light from the event is deflected toward Earth by a sheet of dust in space. The combined travel time of the light from the event to the dust sheet and from there toward Earth is greater than the time taken by the original light that traveled directly toward Earth. This leads to a delay between the original light and its echo.

Luminosity
The amount of light output by an astrophysical object. Also see **Brightness**.

Magnetar
A tightly spinning neutron star with a strong magnetic field, thought to power some types of superluminous supernovae and rapidly evolving transients.

Main sequence
The hydrogen-burning phase of a star's life.

Molecular cloud
A pocket of gas that is dense and large enough for hydrogen molecules to form. Under the pressure of gravity, these clouds collapse and fragment into smaller clouds. Once the temperature in a collapsing cloud is high enough, hydrogen starts fusing into helium, and the cloud becomes a newborn star.

Multimessenger astronomy
A field in astrophysics that uses information gathered from different types of physical phenomena—light, cosmic rays, neutrinos, and gravitational waves— to paint a more complete picture of any astrophysical phenomenon under study.

Nebula
A cloud of gas that either shines on its own (an emission nebula) or obscures the light from background objects (an absorption nebula).

Neutron star
A fast-spinning stellar remnant, created by the core collapse of a star in the range of ~8–25 Solar masses, composed almost entirely of neutrons and kept from collapsing into a black hole by neutron degeneracy pressure (see **Degeneracy pressure**). Neutron stars have masses similar to that of our Sun but are condensed into the size of New York City.

Nova
A thermonuclear explosion on the surface of a white dwarf. Novae are dimmer than supernovae and do not destroy the star. White dwarfs can experience recurrent novae throughout their lifetime.

Nucleosynthesis
The formation of one chemical element out of another.

Outflow
See **Galactic wind**.

Pair-instability supernova
A theoretically predicted type of supernova caused by the collapse of very massive stars. In stars whose masses range between 140 and 260 Solar masses, the collapse of the core leads to the creation of electron-positron pairs, which accelerate the collapse. In stars whose masses range between 100 and 140 Solar masses, it is thought that the pair production mechanism will cause the star to violently pulse several times before it explodes, leading to a "pulsational" pair-instability supernova.

Planetary nebula
The outer envelope of a star less massive than 8 Solar masses, gently blown out into space when the star transitions from a red giant to a white dwarf. Mainly composed of hydrogen, this expanding shell of gas is lit by the ultraviolet light emitted by the white dwarf.

Pulsar
A rapidly spinning neutron star whose polar beams of light periodically sweep across our line of sight, creating a lighthouse effect.

Pulsar-wind nebula
A nebula composed of the light emitted by the spinning cloud of charged particles that are accelerated by the strong magnetic field of a neutron star. See **Nebula**.

Rapidly evolving transient
Any type of astrophysical transient (see **Transient**) whose luminosity rises to a peak in ten days or less and then quickly declines. This supernova class probably contains several distinct types of explosions, each with its own progenitor and explosion physics.

Red giant
A star in the helium-burning phase of its life, directly after the main-sequence stage. In this phase, the star's size expands greatly. Although the star is more luminous at this stage than it was on the main sequence, its surface temperature is lower, giving it a redder appearance.

Redshift
A measure of the velocity at which a light-emitting object is moving away from the observer. The light emitted by the moving object will be stretched in wavelength and appear redder. Light emitted by an object moving toward the observer will be blueshifted. Cosmological redshift is not due to the intrinsic movement of the emitting object (such as stars in galaxies) but to the recession of one galaxy from another due to the expansion of the Universe.

Spectrum
Light, split into its constituent wavelengths by passing through a prism.

Superbubble
A cavity in the space between stars, carved out by supernova explosions and winds from massive stars.

Superluminous supernova
A recently discovered class of supernova, 10 to 100 times more luminous than core-collapse supernovae. It is as yet unclear what powers these supernovae, but suggestions include a large amount of radioactive nickel from the explosion of a very massive star, interaction of the supernova ejecta with a large amount of gas surrounding the star, or a magnetar (see **Magnetar**).

Supernova remnant
A nebula formed by the interaction of the supernova ejecta with the ambient gas surrounding the star before its explosion. Supernova remnants last for thousands of years and shine across the electromagnetic spectrum. However, they are mostly observed in radio and X-ray wavelengths. There are 294 known remnants in our Galaxy and ~1,200 known in other galaxies.

Time-domain astronomy
The study of astrophysical phenomena that change on human timescales (from days to years) instead of cosmic timescales (millions and billions of years). Besides supernovae, this includes variable objects such as pulsating stars, nova eruptions, asteroids, and active supermassive black holes.

Transient
Any astrophysical phenomenon that appears in the night sky and later disappears. This includes supernovae, novae, and other ephemeral phenomena, such as flares from active black holes or tidal disruption events (a flare

produced when a star is shredded by a supermassive black hole). See also **Rapidly evolving transient**.

Type Ia supernova
The thermonuclear explosion of a carbon-oxygen white dwarf. Due to the standardizable nature of their light curves and high luminosities, Type Ia supernovae are widely used to measure distances to other galaxies and probe the nature of dark energy. Yet it is still unclear how white dwarfs are induced to explode. See **White dwarf**.

Wavelength
The distance, in units of length, between two adjacent peaks of a wave. Light, created when electric and magnetic fields are disturbed, is an electromagnetic wave. Different types of light are characterized by different wavelengths: radio and infrared light have longer wavelengths than ultraviolet, X-rays, or gamma rays. Visible light, perceived by our eyes, has wavelengths shorter than infrared but longer than ultraviolet. See figure 7.

White dwarf
The remnant of a star that has stopped fusing elements in its core yet has not exploded as a supernova. Most white dwarfs are composed of carbon and oxygen, but a small fraction of stars more massive than 8 Solar masses are thought to produce white dwarfs made up of oxygen, neon, and either magnesium or iron. White dwarfs are kept from collapsing into neutron stars by electron degeneracy pressure (see **Degeneracy pressure**). A typical white dwarf will have roughly the same mass as our Sun but condensed to the size of Earth. White dwarfs do not produce light; rather, they shine as they cool down from a temperature of over 100,000 Kelvin. Type Ia supernovae are thought to be thermonuclear explosions of white dwarfs.

NOTES

Introduction

1. *Futurama*, "Roswell that Ends Well," first aired 9 December 2001.

2. G. Johns, G. Morrison, G. Rucka, M. Waid, and K. Giffen, *52* (DC Comics, 2007).

3. *Sherlock*, "The Great Game," first aired 8 August 2010.

4. *Star Trek: Voyager*, "The Q and the Grey," first aired 27 November 1996.

5. *Star Trek*, released 8 May 2009; *Star Trek: Picard*, "Remembrance," first aired 23 January 2020.

6. Mass is the amount of matter an object is made of. Weight is mass times the gravitational acceleration of the environment in which the measurement is made. So, while my mass is about 75 kilograms, on the surface of the Earth I weigh 75 kilograms times 9.8 meters per second squared. It is one of the most common misconceptions in physics to refer to mass as weight.

7. Throughout this book, I use the CE ("Common Era") and BCE ("Before the Common Era") year designations.

8. "astronomer, n." and "astrophysicist, n.," *Oxford English Dictionary Online* (December 2019), Oxford University Press.

9. G. E. Hale, "The Astro-physical Journal," *Astronomy and Astro-Physics* 11 (January 1892): 17–22.

10. J. E. Keeler, "The Importance of Astrophysical Research and the Relation of Astrophysics to Other Physical Sciences," *The Astrophysical Journal* 6 (October 1897): 271–288.

Chapter 1

1. Often reductively referred to as the Arab or Muslim Empire, the Islamicate World of the medieval era encompassed the peoples of the Iberian Peninsula, North Africa, the Arabian Peninsula, Persia, and Central Asia. With Arabic as a shared language of science and religion, Islamicate scholars of various religions and cultural backgrounds made great strides in many fields of study, including mathematics, astronomy, and medicine.

2. D. A. Green and W. Orchiston, "In Search of *Mahutonga*: A Possible Supernova Recorded in Maori Astronomical Traditions?," *Archaeoastronomy* 18 (2004): 110–113.

3. D. W. Hamacher, "Are Supernovae Recorded in Indigenous Astronomical Traditions?," *Journal of Astronomical History and Heritage* 17, no. 2 (July 2014): 161–170.

4. J. C. Brandt and R. A. Williamson, "The 1054 Supernova and Native American Rock Art," *Journal for the History of Astronomy* 10 (1979): S1–S38.

5. F. R. Stephenson, "Historical Records of Supernovae," in *Handbook of Supernovae*, ed. A. W. Alsabti and P. Murdin (Springer, 2017), 49–62.

6. *Nanmen*, the Southern Gate, was one of ~300 constellations used by Chinese astronomers.

7. F. R. Stephenson and D. A. Green, *Historical Supernovae and Their Remnants* (Oxford University Press, 2002), 187–188; modified by the author.

8. R. Stothers, "Is the Supernova of A.D. 185 Recorded in Ancient Roman Literature?," *Isis* 68 (September 1977): 443–447.

9. Herodianus, *Historia de imperio post Marcum*, 1.14.1, ca. 250 CE, as translated by E. C. Nichols, *Herodian of Antioch's History of the Roman Empire* (University of California Press, 1961).

10. Unknown, *Historia Augusta*, 16.1–2, fourth century CE, as modified from the translation of D. Magie, *The Scriptores Historiae Augustae* (Harvard University Press, 1921). The "war of the deserters" may refer to the military campaign against deserters from the Roman armies in Gaul in 186 CE.

11. Stephenson and Green, *Historical Supernovae and Their Remnants*, 151–162.

12. B. R. Goldstein, "Evidence for a Supernova of A.D. 1006," *The Astronomical Journal* 70, no. 1 (February 1965): 105–114.

13. Ibid.

14. Modified from Stephenson and Green, *Historical Supernovae and Their Remnants*, 168.

15. Modified from ibid., 169.

16. F. F. Gardner and D. K. Milne, "The Supernova of A.D. 1006," *The Astronomical Journal* 70, no. 9 (November 1965): 754.

17. D. Gans, *Sefer Zemah David* (Prague: Shlomo Cohen and his brother's son Moshe Cohen Press, 1592), as translated by the author. The year 332, as Gans writes it, refers to the year 5332 by the Jewish count. I have chosen to translate "התחדש ונראה" as "seen anew" since the modern meaning of "התחדש" is "to be renewed / made new." This choice makes sense in the context of a returning comet. The final clause of this excerpt, "God in his mercy frustrates the signs of liars," is a fragment from Isaiah, 44:25. The English Standard Version of the Bible, which provides a more literal translation than the King James version, translates this passage as: "who frustrates the signs of liars and makes fools of diviners, who turns wise men back and makes their knowledge foolish." A previous translation of this excerpt, by Dr. David Cook, appears in Stephenson and Green, *Historical Supernovae and Their Remnants*, 93. However, that translation omits the mention of the star's large tail and the allusion to Isaiah 44:25.

18. This view is supported by M. Breuer, who notes that one of the many chronicles Gans relied on (H. Goltz, *Keyserliche Chronik*, Frankfurt am Main 1588) mentioned that the new star was first mistaken for a comet. See footnote 70 in M. Breuer, *Zemah David: A Chronicle of Jewish and World History* (Jerusalem: Magnes Press, 1983), 405–406.

19. As also noted by Stephenson and Green, *Historical Supernovae and Their Remnants*, 93.

20. Babylonian Talmud: Tractate Baba Bathra, 12b, as translated by the author. The "Temple" refers to the First Temple in Jerusalem, traditionally built by King Solomon and razed by the Babylonians during the sixth century BCE.

21. W. Baade, "B Cassiopeiae as a Supernova of Type I," *The Astrophysical Journal* 102 (November 1945): 309–317; P. Ruiz-Lapuente, "Tycho Brahe's Supernova: Light from Centuries Past," *The Astrophysical Journal* 612, no. 1 (September 2004): 357–363.

22. J. H. Walden, "Tycho Brahe: On a New Star, Not Previously Seen within the Memory of Any Age since the Beginning of the World," in H. Shapley and H. E. Howarth, *A Source Book in Astronomy* (McGraw-Hill, 1929), 13–20.

23. G. Galilei, *Sidereus nuncius* (Venice: apud Thomam Baglionum, 1610).

Chapter 2

1. To be precise, reflecting telescopes can have shorter relative focal lengths (also called *f* numbers), which makes them more compact.

2. K. Lundmark, "The Pre-Tychonic Novae," *Lund Observatory Circular* 8 (December 1932): 216–218. In this paper, Lundmark estimates a high average brightness for several historical nova candidates, which in his words "places the objects among the 'upper class' or *super-Novae*." The italics appear in the original, lending credence to the assumption that Lundmark came up with the term. In the same paper, he also suggested calling the historical novae *pre-Tychonic*, again in italics. The latter term did not catch on.

3. This assertion is also backed up by the *Oxford English Dictionary*.

4. E. Hartwig et al., "Ueber eine Veränderung des grossen Andromedanebels," *Astronomische Nachrichten* 112 (September 1885): 245–248; M. Wolf, "Var. 6. 1909 Ursae majoris," *Astronomische Nachrichten* 180 (March 1909): 245–246.

5. For biographies of Walter Baade and Fritz Zwicky, see D. E. Osterbrock, *Walter Baade: A Life in Astrophysics* (Princeton: Princeton University Press, 2001); A. Stöckli and R. Müller, *Fritz Zwicky: An Extraordinary Astrophysicist* (Cambridge: Cambridge Scientific Publishers, 2016); and J. Johnson, Jr., *Zwicky: The Outcast Genius Who Unmasked the Universe* (Cambridge, MA: Harvard University Press, 2019).

6. W. Baade and F. Zwicky, "On Super-Novae," *Proceedings of the National Academy of Sciences* 20, no. 5 (March 1934): 254–259.

7. J. Chadwick, "Possible Existence of a Neutron," *Nature* 129, no. 3252 (February 1932): 312.

8. W. Baade and F. Zwicky, "Cosmic Rays from Super-Novae," *Proceedings of the National Academy of Sciences* 20, no. 5 (March 1934): 259–263.

9. S. Starrfield et al., "CNO Abundances and Hydrodynamic Models of the Nova Outburst," *The Astrophysical Journal* 176 (August 1972): 169–176; D. Prialnik, M. M. Shara, and G. Shaviv, "The Evolution of a Slow Nova Model with a $Z = 0.03$ Envelope from Pre-explosion to Extinction," *Astronomy & Astrophysics* 62, no. 3 (January 1978): 339–348.

10. F. Zwicky, "Basic Results of the International Search for Supernovae," *Annales d'Astrophysique* 27, no. 4 (May 1964): 300–312. The astronomers who took part in the first international survey were Max Schürer and Paul Wild at the Zimmerwald Observatory in Bern, Switzerland; Giuliano Romano and Leonida Rosino at the Asiago Astrophysical Observatory, Italy; and Guillermo Haro and Enrique Chavira at the Tonantzintla Observatory, Mexico. Once the Committee for Research on Supernovae was created, they were joined by Charles Bertaud at Meudon Observatory, France; Edwin F. Carpenter at the Steward Observatory in Tucson, Arizona; Boris V. Kukarkin at the Moscow University Observatory, USSR; and José Luis Sérsic at Córdoba Observatory, Argentina.

11. The larger the radius, r, of a telescope's mirror, the more surface area, πr^2, is available for collecting photons.

12. R. P. Kirshner, personal communication, 20 May 2019. As a graduate student at Caltech, Kirshner took part in Zwicky's supernova survey during 1971.

13. Ibid.

14. F. Zwicky, "On the Frequency of Supernovae," *The Astrophysical Journal* 88, no. 5 (December 1938): 529–541.

15. W. S. Boyle and G. E. Smith, "Charge Coupled Semiconductor Devices," *Bell System Technical Journal* 49, no. 4 (April 1970): 587–593.

16. The first use of the term "magnitude" is attributed to the second-century CE scholar Claudius Ptolemaeus (commonly referred to as Ptolemy), who described the brightness of the stars listed in his *Almagest* on a scale ranging from the "first magnitude" (brightest) to the "sixth magnitude" (faintest). R. Miles, "A Light History of Photometry: From Hipparchus to the Hubble Space Telescope," *Journal of the British Astronomical Association* 117, no. 4 (August 2007): 172–186.

17. J. Leaman et al., "Nearby Supernova Rates from the Lick Observatory Supernova Search—I. The Methods and Data Base," *Monthly Notices of the Royal Astronomical Society* 412, no. 3 (April 2011): 1419–1440.

18. J. J. Johnson, "A Supernova in NGC 4725," *Publications of the Astronomical Society of the Pacific* 52, no. 307 (June 1940): 206.

19. Not all supernovae receive an official designation, such as SN 1938A. BUF-19Sop and BUF19Awd are internal designations used by the Beyond Ultradeep Frontier Fields and Legacy Observations (BUFFALO) supernova survey. They denote the name of the survey (BUF), the year in which the supernovae were discovered (2019), and the nicknames given them by the searchers. In this case, we named them after my daughters, Sophia and Awdrey.

Chapter 3

1. The energy of a wave, E, is directly proportional to its frequency, ν, or indirectly proportional to its wavelength, λ, as $E = h\nu = hc/\lambda$, where h is Planck's constant and c is the speed of light.

2. R. Minkowski, "Spectra of Supernovae," *Publications of the Astronomical Society of the Pacific* 53, no. 314 (August 1941): 224–225.

3. Fritz Zwicky's Types III, IV, and V did not make the cut. F. Zwicky, "Basic Results of the International Search for Supernovae," *Annales d'Astrophysique* 27, no. 4 (May 1964): 300–312.

4. For a review of the basic concepts behind the modern classification scheme, see A. Gal-Yam, "Observational and Physical Classification of Supernovae," in *Handbook of Supernovae* (Springer, 2017), 195–237. See also A. V. Filippenko, "Optical Spectra of Supernovae," *Annual Review of Astronomy and Astrophysics* 35 (September 1997): 309–355.

5. Sources for figure 9: **Ia**, U. Munari et al., "BVRI Lightcurves of Supernovae SN 2011fe in M101, SN 2012aw in M95, and SN 2012cg in NGC 4424," *New Astronomy* 20 (April 2013): 30–37; **Ib**, S. Srivastav et al., "Optical Observations of the Fast Declining Type Ib Supernova iPTF13bvn," *Monthly Notices of the Royal Astronomical Society* 445, no. 2 (December 2014): 1932–1941; **Ic**, R. J. Foley et al., "Optical Photometry and Spectroscopy of the SN 1998bw-like Type Ic Supernova 2002ap," *Publications of the Astronomical Society of the Pacific* 115, no. 812 (October 2003): 1220–1235; **IIP**, M. Dall'Ora et al., "The Type IIP Supernova 2012aw in M95: Hydrodynamical Modeling of the Photospheric Phase from Accurate Spectrophotometric Monitoring," *The Astrophysical Journal* 787, no. 2 (June 2014): 139; **IIL**, S. Bose et al., "Photometric and Polarimetric Observations of Fast Declining Type II Supernovae 2013hj and 2014G," *Monthly Notices of the Royal Astronomical Society* 455, no. 3 (January

2016): 2712–2730; **II-87A**, J. W. Menzies et al., "Spectroscopic and Photometric Observations of SN 1987A: The First 50 Days," *Monthly Notices of the Royal Astronomical Society* 227 (August 1987): 39P–49P; R. M. Catchpole et al., "Spectroscopic and Photometric Observations of SN 1987A—II. Days 51 to 134," *Monthly Notices of the Royal Astronomical Society* 229 (November 1987): 15P–25P; and R. M. Catchpole et al., "Spectroscopic and Photometric Observations of SN 1987A—III. Days 135 to 260," *Monthly Notices of the Royal Astronomical Society* 231 (April 1988): 75P–89P; **IIn**, A. Pastorello et al., "Interacting Supernovae and Supernova Impostors: SN 2009ip, Is This the End?," *The Astrophysical Journal* 767, no. 1 (April 2013); 1; and M. Fraser et al., "SN 2009ip à la PESSTO: No Evidence for Core Collapse Yet," *Monthly Notices of the Royal Astronomical Society* 433, no. 2 (August 2013): 1312–1337.

6. Sources for figure 10: **IIP**, D. C. Leonard et al., "The Distance to SN 1999em in NGC 1637 from the Expanding Photosphere Method," *Publications of the Astronomical Society of the Pacific* 114, no. 791 (January 2002): 35–64; **IIn**, N. Smith et al., "Systematic Blueshift of Line Profiles in the Type IIn Supernova 2010jl: Evidence for Post-shock Dust Formation?," *The Astronomical Journal* 143, no. 1 (January 2012): 17; **Ib**, Y. Cao et al., "Discovery, Progenitor and Early Evolution of a Stripped Envelope Supernova iPTF13bvn," *The Astrophysical Journal Letters* 775, no. 1 (September 2013), L7; **Ic**, S. Valenti et al., "The Carbon-Rich Type Ic SN 2007gr: The Photospheric Phase," *The Astrophysical Journal Letters* 673, no. 2 (February 2008): L155; **Ia**, P. A. Mazzali et al., "Hubble Space Telescope Spectra of the Type Ia Supernova SN 2011fe: A Tail of Low-Density, High-Velocity Material with $Z < Z_\odot$," *Monthly Notices of the Royal Astronomical Society* 439, no. 2 (April 2014): 1959–1979.

7. For example, there are ongoing debates over whether Type IIP and IIL supernovae both originate from the core collapse of single, 8–25-Solar-mass stars, and whether 1991bg-like, normal, and 1991T-like Type Ia supernovae all exist along a continuum. See I. Arcavi et al., "Caltech Core-Collapse Project (CCCP) Observations of Type II Supernovae: Evidence for Three Distinct Photometric Subtypes," *The Astrophysical Journal Letters* 756, no. 2 (September 2012): L30; T. Faran et al., "A Sample of Type II-L Supernovae," *Monthly Notices of the Royal Astronomical Society* 445, no. 1 (November 2014): 554–569; and E. Heringer et al., "Spectral Sequences of Type Ia Supernovae. I. Connecting Normal and Subluminous SNe Ia and the Presence of Unburned Carbon," *The Astrophysical Journal* 846, no. 1 (September 2017): 15.

8. Usually ascribed to Type II and Ib/c supernovae, shock breakouts might also occur in Type Ia supernovae. See A. L. Piro et al., "Shock Breakout from Type Ia Supernovae," *The Astrophysical Journal* 708, no. 1 (January 2010): 598–604.

9. E. Waxman and B. Katz, "Shock Breakout Theory," in *Handbook of Supernovae*, 967–1015; P. M. Garnavich et al., "Shock Breakout and Early Light Curves of Type-IIP Supernovae Observed with Kepler," *The Astrophysical Journal* 820, no. 1 (March 2016): 23; M. C. Bersten et al., "A Surge of Light at the Birth of a Supernova," *Nature* 554, no. 7693 (February 2018): 497–499.

10. First noted by T. Pankey, Jr., "Possible Thermonuclear Activities in Natural Terrestrial Minerals," Ph.D. thesis, Howard University, 1962. The two papers usually cited on this issue are S. A. Colgate and C. McKee, "Early Supernova Luminosity," *The Astrophysical Journal* 157 (August 1969): 623–662, and J. W. Truran et al., "Nucleosynthesis in Supernova Shock Waves," *Canadian Journal of Physics* 45, no. 7 (July 1967): 2315–2332.

11. O. Graur et al., "A Year-Long Plateau in the Late-Time Near-Infrared Light Curves of Type Ia Supernovae," *Nature Astronomy* 4 (February 2020): 188–195. The radioactive decays of ^{56}Ni and ^{56}Co proceed through the capture of an electron or ejection of a positron (inverse beta decay, also called the "*e*-process"). The daughter nucleus from this nuclear reaction will be in an energetically excited state and will decay to the ground state by emitting high-energy photons in the gamma-ray band. The atoms that make up the expanding supernova ejecta absorb these photons, which raises them to higher energy levels. These atoms also de-excite by emitting photons, but these new photons will have different, lower energies than the original gamma rays. The photons that finally make their way out of the ejecta will have a range of energies in the optical and near-infrared bands of the electromagnetic spectrum. This process of conversion from gamma rays to optical/near-infrared is called "thermalization."

12. R. A. Chevalier, "The Interaction of Supernovae with the Interstellar Medium," *Annual Review of Astronomy and Astrophysics* 15 (1977): 175–196.

13. P. A. Mazzali et al., "A Common Explosion Mechanism for Type Ia Supernovae," *Science* 315, no. 5813 (February 2007): 825–828.

14. D. Maoz and A. Gal-Yam, "The Type Ia Supernova Rate in $z<=1$ Galaxy Clusters: Implications for Progenitors and the Source of Cluster Iron," *Monthly Notices of the Royal Astronomical Society* 347, no. 3 (January 2004): 951–956.

15. Calculated as $R(z) = \int (R_{Ia}(z) + R_{CC}(z)) dV(z)$, where $R_{Ia}(z)$ is the Type Ia supernova rate from D. Maoz and O. Graur, "Star Formation, Supernovae, Iron, and α: Consistent Cosmic and Galactic Histories," *The Astrophysical Journal* 848, no. 1 (October 2017): 25; $R_{CC}(z)$ is the core-collapse rate, represented by the cosmic star-formation history scaled to the present-day core-collapse supernova rate measured by S. Mattila et al., "Core-Collapse Supernovae Missed by Optical Surveys," *The Astrophysical Journal* 756, no. 2 (September 2012): 111. The cosmic star-formation history was taken from P. Madau and T. Fragos,

"Radiation Backgrounds at Cosmic Dawn: X-Rays from Compact Binaries," *The Astrophysical Journal* 840, no. 1 (May 2017): 39; and the comoving volume, $V(z) = (4\pi/3)[D_L/(1 + z)]^3$, is calculated from the luminosity distance, D_L, assuming a Hubble constant value of $H_0 = 70$ km/s/Mpc, matter density $\Omega_m = 0.3$, and dark-energy density $\Omega_\Lambda = 0.7$. The integral is evaluated from redshifts $z = 0$ to 8. Note that this rate is representative of the "visible Universe," i.e., the part of the Universe from which light has had enough time to reach our telescopes. The size of the full Universe is still unknown.

16. L.-G. Strolger et al., "The Rate of Core Collapse Supernovae to Redshift 2.5 from the CANDELS and CLASH Supernova Surveys," *The Astrophysical Journal* 813, no. 2 (November 2015): 93; Maoz and Graur, "Star Formation, Supernovae, Iron, and α," 25.

17. O. Graur et al., "LOSS Revisited. II. The Relative Rates of Different Types of Supernovae Vary between Low- and High-Mass Galaxies," *The Astrophysical Journal* 837, no. 2 (March 2017): 121.

18. SNe 1917A, 1939C, 1948B, 1968D, 1969P, 1980K, 2002hh, 2004et, 2008S, and 2017eaw.

19. W. Li et al., "Nearby Supernova Rates from the Lick Observatory Supernova Search—III. The Rate-Size Relation, and the Rates as a Function of Galaxy Hubble Type and Colour," *Monthly Notices of the Royal Astronomical Society* 412, no. 3 (April 2011): 1473–1507.

20. R. A. Fesen, "The Expansion Asymmetry and Age of the Cassiopeia A Supernova Remnant," *The Astrophysical Journal* 645, no. 1 (July 2006): 283–292; S. P. Reynolds et al., "The Youngest Galactic Supernova Remnant: G1.9+0.3," *The Astrophysical Journal Letters* 680, no. 1 (June 2008): L41.

Chapter 4

1. The photons emitted during the fusion process in the star's core also exert an outward pressure (called radiation pressure). In most stars, the pressure exerted by the motion of the gas dominates over the radiation pressure exerted by the photons. This changes in massive stars, where the internal temperature is high enough for radiation pressure to dominate.

2. Photons created by the fusion process have to make their way out through the star. On their way, they are absorbed by the atoms of hydrogen gas they encounter, which then emit new photons in random directions. This process can be described as a "random walk" of one photon as it makes its way through the star. In our Sun, for example, this random walk lasts ~50,000 years until a photon finally escapes.

3. This state is called "hydrostatic equilibrium."

4. In more technical terms, the distribution of wavelengths of all the emerging photons is called a "black-body" distribution. Similar to a bell curve, it rises to a peak and then declines. The hotter the star, the shorter the wavelength at which the black-body distribution peaks.

5. According to the Pauli exclusion principle, two fermions—elementary particles with half-integer spins, such as electrons and neutrons—cannot have the same quantum numbers and simultaneously occupy the same quantum state. For example, each orbital in an atom, which can also be described as a distinct energy state, can be occupied by two electrons, but those electrons must have opposite spins: ½ and –½. As the gas in the star's core is condensed under gravitational pressure, electrons will fill up the available energy states and start to resist the addition of other electrons to their energy states, creating a new outward-facing source of pressure. Quantum degeneracy pressure is responsible for the stability of white dwarfs (via electrons) and neutron stars (via neutrons) against gravitational collapse.

6. The term "planetary nebula" was coined by William Herschel in the eighteenth century. Through his relatively weak telescopes, these objects looked round, like the planets. M. Hoskin, "William Herschel and the Planetary Nebulae," *Journal for the History of Astronomy* 45, no. 2 (May 2014): 209–225.

7. S. E. Woosley, A. Heger, and T. A. Weaver, "The Evolution and Explosion of Massive Stars," *Reviews of Modern Physics* 74, no. 4 (November 2002): 1015–1071.

8. First, the iron atoms in the core begin to absorb photons, which break the iron atoms into helium atoms and neutrons. The helium atoms also absorb photons and—through the same process of "photodisintegration"—break apart into their constituent protons and neutrons. Second, the enormous pressure in the core allows protons to merge with electrons and transform into neutrons. This process of "neutronization" lands three blows on the stability of the core: it consumes electrons, which removes a source of pressure against gravity; requires energy, which is taken from the energy used to gravitationally bind the core together; and produces neutrinos that carry away energy.

9. The interaction of the supernova ejecta with the surrounding gas is "collisionless," i.e., the particles that make up the ejecta and the gas do not interact with each other directly. Instead, the interaction (the shock) is caused by the electromagnetic field created by the hot plasma that comprises the ejecta.

10. Referred to as the "circumstellar medium" and "interstellar medium," respectively.

11. D. A. Green, "A Catalogue of 294 Galactic Supernova Remnants," *Bulletin of the Astronomical Society of India* 42, no. 2 (June 2014): 47–58; G. Dubner and

E. Giacani, "Radio Emission from Supernova Remnants," *The Astronomy and Astrophysics Review* 23 (September 2015): 3.

12. To be more precise, during this so-called "Sedov-Taylor" phase, the remnant can be described as a ball filled with two fluids: an outer layer made up of the swept-up shocked gas lying on top of an inner layer composed of the dense supernova ejecta. This type of structure—a dense fluid under a thinner fluid—leads to the formation of Rayleigh-Taylor instabilities, which manifest as "dimples" and curling "fingers" at the contact surface between the two fluids.

13. Specifically, the speed of sound in the interstellar medium.

14. Black holes created by core-collapse or pair-instability supernova explosions do not radiate and are invisible to telescopes. The existence of black holes in the centers of galaxies has been inferred from the movements of stars orbiting them or by directly imaging their event horizons. However, these are all supermassive black holes with masses more than a million times the mass of our Sun. Stellar-mass black holes created by supernova explosions are expected to have masses of just a few to tens of Solar masses. Binaries composed of black holes weighing a few tens of Solar masses each have been detected through the gravitational waves they emit when they merge (chapter 8).

Chapter 5

1. R. A. Alpher, H. Bethe, and G. Gamow, "The Origin of Chemical Elements," *Physical Review* 73, no. 7 (April 1948): 803–804.

2. This is called the "x-process" or "cosmic-ray spallation."

3. E. M. Burbidge, G. R. Burbidge, W. A. Fowler, and F. Hoyle, "Synthesis of the Elements in Stars," *Reviews of Modern Physics* 29, no. 4 (October 1957): 547–650.

4. Hydrogen fuses into helium via two processes. In the *p-p chain*, two hydrogen atoms (that is, two protons) merge to form a deuterium atom (one proton and one neutron). The deuterium atom fuses with a hydrogen atom to form ^{3}He, composed of two protons and a neutron. Finally, two ^{3}He atoms merge to form ^{4}He—stable helium—composed of two protons and two neutrons. In stars more massive than 1.2 Solar masses, the dominant process is the *CNO cycle*, in which existing traces of carbon, nitrogen, and oxygen are used as stepping stones between hydrogen and helium.

5. The original iron created during the silicon-burning stage becomes trapped in the neutron star or black hole that forms out of the star's collapsed core.

6. For a detailed description of the nucleosynthesis processes that take place at each burning stage, consult the following sources. For massive stars: S. E. Woosley, A. Heger, and T. A. Weaver, "The Evolution and Explosion of Massive

Stars," *Reviews of Modern Physics* 74, no. 4 (October 2002): 1015–1071. For white-dwarf explosions: I. R. Seitenzahl and D. M. Townsley, "Nucleosynthesis in Thermonuclear Supernovae," in *Handbook of Supernovae* (Springer, 2017), 1955–1978.

7. R. Hirschi, "Pre-supernova Evolution and Nucleosynthesis in Massive Stars and Their Stellar Wind Contribution," in *Handbook of Supernovae*, 1879–1929.

8. The shock's velocity should be 20–45 kilometers per second. H. A. T. Vanhala and A. G. W. Cameron, "Numerical Simulations of Triggered Star Formation. I. Collapse of Dense Molecular Cloud Cores," *The Astrophysical Journal* 508, no. 1 (November 1998): 291–307; K. M. Desai et al., "Supernova Remnants and Star Formation in the Large Magellanic Cloud," *The Astronomical Journal* 140, no. 2 (August 2010): 584–594.

9. J. R. Dawson et al., "Supergiant Shells and Molecular Cloud Formation in the Large Magellanic Cloud," *The Astrophysical Journal* 763, no. 1 (January 2013): 56; S. Ehlerová and J. Palouš, "HI Shells in the Leiden/Argentina/Bonn HI survey," *Astronomy & Astrophysics* 550 (February 2013): A23.

10. The space between galaxies is known as the "intergalactic medium."

11. T. M. Heckman, M. D. Lehnert, and L. Armus, "Galactic Superwinds," in *The Environment and Evolution of Galaxies* (Springer, 1993); T. Naab and J. P. Ostriker, "Theoretical Challenges in Galaxy Formation," *Annual Review of Astronomy and Astrophysics* 55, no. 1 (August 2017): 59–109.

12. K. Kurz, "Die radioaktiven Stoffe in Erde und Luft als Ursache der durchdringenden Strahlung in der Atmosphäre," *Physikalische Zeitschrift* 10 (October 1909): 834–845.

13. V. Hess, "Über Beobachtungen der durchdringenden Strahlung bei sieben Freiballonfahrten," *Physikalische Zeitschrift* 13 (November 1912): 1084–1091.

14. Enrico Fermi first suggested that cosmic rays could be accelerated by turbulent magnetic fields: E. Fermi, "On the Origin of the Cosmic Radiation," *Physical Review* 75, no. 8 (April 1949): 1169–1174. The connection to shockwaves was made by W. I. Axford, E. Leer, and G. Skadron, "The Acceleration of Cosmic Rays by Shock Waves," in *Proceedings of the 15th International Cosmic Ray Conference* 11 (1977): 132. For a review, see G. Morlino, "High-Energy Cosmic Rays from Supernovae," in *Handbook of Supernovae*, 1711–1736.

15. F. A. Cucinotta and E. Cacao, "Non-Targeted Effects Models Predict Significantly Higher Mars Mission Cancer Risk than Targeted Effects Models," *Scientific Reports* 7 (May 2017): 1832.

16. J. Maíz-Apellániz, "The Origin of the Local Bubble," *The Astrophysical Journal Letters* 560, no. 1 (October 2001): L83–L86; N. Benítez, J. Maíz-Apellániz, and M. Canelles, "Evidence for Nearby Supernova Explosions," *Physical Review Letters* 88, no. 8 (February 2002): 081101.

17. S. Portegies Zwart et al., "The Consequences of a Nearby Supernova on the Early Solar System," *Astronomy & Astrophysics* 616 (August 2018): A85.

18. Half-life, the time it takes half the mass of a given radioactive element to decay into other elements, is a measure of that element's radioactivity: more radioactive elements will have shorter half-lives than more stable elements.

19. K. Knie et al., "Indication for Supernova Produced [60]Fe Activity on Earth," *Physical Review Letters* 83, no. 1 (July 1999): 18–21; K. Knie et al., "[60]Fe Anomaly in a Deep-Sea Manganese Crust and Implications for a Nearby Supernova Source," *Physical Review Letters* 93, no. 17 (October 2004): 171103; C. Fitoussi et al., "Search for Supernova-Produced [60]Fe in a Marine Sediment," *Physical Review Letters* 101, no. 12 (September 2008): 121101; A. Wallner et al., "Recent Near-Earth Supernovae Probed by Global Deposition of Interstellar Radioactive [60]Fe," *Nature* 532, no. 7597 (April 2016): 69–72 (the authors found [60]Fe signals in two time bins: 1.5–3.2 and 6.5–8.7 million years ago, indicating there may have been two distinct periods of supernova activity near the Solar System in the last 10 million years); P. Ludwig et al., "Time-Resolved 2-Million-Year-Old Supernova Activity Discovered in Earth's Microfossil Record," *Proceedings of the National Academy of Sciences* 113, no. 33 (August 2016): 9232–9237.

20. L. Fimiani et al., "Interstellar [60]Fe on the Surface of the Moon," *Physical Review Letters* 116, no. 15 (April 2016): 151104.

21. M. Kachelrieß, A. Neronov, and D. V. Semikoz, "Signatures of a Two Million Year Old Supernova in the Spectra of Cosmic Ray Protons, Antiprotons, and Positrons," *Physical Review Letters* 115, no. 18 (October 2015): 181103.

22. Specifically, a third of all genera of marine megafauna (i.e., large animals). Various species of small cat including the domestic cat, for example, make up the genus (singular of genera) *Felis*, which is part of the Felidae family.

23. C. Pimiento et al., "The Pliocene Marine Megafauna Extinction and Its Impact on Functional Diversity," *Nature Ecology & Evolution* 1 (August 2017): 1100–1106.

24. N. Gehrels et al., "Ozone Depletion from Nearby Supernovae," *The Astrophysical Journal* 585, no. 2 (March 2003): 1169–1176.

25. M. Beech, "The Past, Present and Future Supernova Threat to Earth's Biosphere," *Astrophysics and Space Science* 336, no. 2 (December 2011): 287–302.

Chapter 6

1. For information about how supernovae are used to test time dilation and the variability of physical constants, see S. Blondin et al., "Time Dilation in Type Ia Supernova Spectra at High Redshift," *The Astrophysical Journal* 682, no. 2 (August 2008): 724–736; A. Ferrero and B. Altschul, "Limits on the Time

Variation of the Fermi Constant G_F Based on Type Ia Supernova Observations," *Physical Review D* 82, no. 12 (December 2010): 123002; and J. Mould and S. A. Uddin, "Constraining a Possible Variation of G with Type Ia Supernovae," *Publications of the Astronomical Society of Australia* 31 (March 2014): e015.

2. Paraphrasing E. Idle and J. Du Prez, "Galaxy Song," in *Monty Python's Meaning of Life* (1983); and R. Rogel, "Yakko's Universe," in *Animaniacs*, Episode 3 (aired 15 September 1993).

3. H. S. Leavitt and E. C. Pickering, "Periods of 25 Variable Stars in the Small Magellanic Cloud," *Harvard College Observatory Circular* 173 (March 1912): 1–3. The life and work of Henrietta Leavitt are recounted in two recent biographies: G. Johnson, *Miss Leavitt's Stars* (New York: W. W. Norton, 2005); and D. Sobel, *The Glass Universe* (New York: Viking Penguin, 2016).

4. B. W. Rust, "The Use of Supernovae Light Curves for Testing the Expansion Hypothesis and Other Cosmological Relations" (Ph.D. thesis, University of Illinois at Urbana-Champaign, 1974); Yu. P. Pskovskii, "Light Curves, Color Curves, and Expansion Velocity of Type I Supernovae as Functions of the Rate of Brightness Decline," *Soviet Astronomy* 21, no. 6 (December 1977): 675–682.

5. M. M. Phillips, "The Absolute Magnitudes of Type Ia Supernovae," *The Astrophysical Journal Letters* 413 (August 1993): L105–L108.

6. Most contemporary astronomers, including myself, commonly refer to this correlation as "the Phillips relation," without referring to either Rust or Pskovskii. This omission is accidental but reflects a wider pattern in science where, over time, discoveries are often attributed to one person and independent discoverers of the same phenomenon are sidelined or forgotten. Some other examples include the Auger effect, which was also independently discovered by Lise Meitner, and the Kozai effect, which is now correctly referred to as the Lidov-Kozai effect.

7. M. Hamuy and P. A. Pinto, "Type II Supernovae as Standardized Candles," *The Astrophysical Journal Letters* 566, no. 2 (February 2002): L63–L65; C. Inserra and S. J. Smartt, "Superluminous Supernovae as Standardizable Candles and High-Redshift Distance Probes," *The Astrophysical Journal* 796, no. 2 (December 2014): 87.

8. A. G. Riess et al., "Observational Evidence from Supernovae for an Accelerating Universe and a Cosmological Constant," *The Astronomical Journal* 116, no. 3 (September 1998): 1009–1038; S. Perlmutter et al., "Measurements of Ω and Λ from 42 High-Redshift Supernovae," *The Astrophysical Journal* 517, no. 2 (June 1999): 565–586.

9. The term "dark energy" is attributed to Michael S. Turner: S. Pelmutter, M. S. Turner, and M. White, "Constraining Dark Energy with Type Ia Supernovae

and Large-Scale Structure," *Physical Review Letters* 83, no. 4 (July 1999): 670–673.

10. A. Einstein, "Kosmologische Betrachtungen zur allgemeinen Relativitätstheorie," *Sitzungsberichte der Königlich Preußischen Akademie der Wissenschaften* 6 (February 1917): 142–152.

11. E. Hubble, "A Relation between Distance and Radial Velocity among Extra-Galactic Nebulae," *Proceedings of the National Academy of Sciences* 15, no. 3 (March 1929): 168–173.

12. Ya. B. Zel'dovich, "Special Issue: The Cosmological Constant and the Theory of Elementary Particles," *Soviet Physics Uspekhi* 11, no. 3 (March 1968): 381–393.

13. J. A. Frieman, M. S. Turner, and D. Huterer, "Dark Energy and the Accelerating Universe," *Annual Review of Astronomy and Astrophysics* 46 (June 2008): 385–432.

14. Specifically, galaxies in the "Hubble flow," i.e., at redshifts between ~0.03 and ~0.1. These galaxies are far enough that their redshifts are mostly due to the expansion of the Universe and not to their own movement through space, yet close enough that the result is not strongly dependent on other cosmological parameters.

15. The recession velocity of a galaxy, v, is connected to its redshift, z, by the speed of light, c, so that $v = cz$. Hubble's law states that $v = H_0 D$, where D is the distance to the galaxy, so $cz = H_0 D$.

16. A. G. Riess et al., "A 2.4% Determination of the Local Value of the Hubble Constant," *The Astrophysical Journal* 826, no. 1 (July 2016): 56; A. G. Riess et al., "Cosmic Distances Calibrated to 1% Precision with Gaia EDR3 Parallaxes and Hubble Space Telescope Photometry of 75 Milky Way Cepheids Confirm Tension with ΛCDM," *The Astrophysical Journal Letters* 908, no. 1 (February 2021): L6.

17. F. Zwicky, "Die Rotverschiebung von extragalaktischen Nebeln," *Helvetica Physica Acta* 6 (November 1933): 110–127.

18. V. C. Rubin and W. K. Ford, Jr., "Rotation of the Andromeda Nebula from a Spectroscopic Survey of Emission Regions," *The Astrophysical Journal* 159 (February 1970): 379–403.

19. S. Refsdal, "On the Possibility of Determining Hubble's Parameter and the Masses of Galaxies from the Gravitational Lens Effect," *Monthly Notices of the Royal Astronomical Society* 128 (January 1964): 307–310.

20. P. L. Kelly et al., "Multiple Images of a Highly Magnified Supernova Formed by an Early-Type Cluster Galaxy Lens," *Science* 347, no. 6226 (March 2015): 1123–1126.

21. P. L. Kelly et al., "Déjà vu All Over Again: The Reappearance of Supernova Refsdal," *The Astrophysical Journal Letters* 819, no. 1 (March 2016): L8.

22. B. J. Williams and T. Temim, "Infrared Emission from Supernova Remnants: Formation and Destruction of Dust," in *Handbook of Supernovae* (Springer, 2017), 2105–2124.

23. F. Patat et al., "Studying the Small Scale ISM Structure with Supernovae," *Astronomy & Astrophysics* 514 (May 2010): A78.

24. F. Patat, "Reflections on Reflexions—I. Light Echoes in Type Ia Supernovae," *Monthly Notices of the Royal Astronomical Society* 357, no. 4 (March 2005): 1161–1177; Y. Yang et al., "Interstellar-Medium Mapping in M82 through Light Echoes around Supernova 2014J," *The Astrophysical Journal* 834, no. 1 (January 2017): 60.

Chapter 7

1. S. D. Van Dyk, "Supernova Progenitors Observed with *HST*," in *Handbook of Supernovae* (Springer, 2017), 693–719.

2. S. J. Smartt, "Progenitors of Core-Collapse Supernovae," *Annual Review of Astronomy and Astrophysics* 47, no. 1 (September 2009): 63–106 and references therein.

3. G. L. White and D. F. Malin, "Possible Binary Star Progenitor for SN1987A," *Nature* 327, no. 6117 (May 1987): 36–38.

4. G. Aldering et al., "SN 1993J: The Optical Properties of Its Progenitor," *The Astronomical Journal* 107 (February 1994): 662–672; J. R. Maund et al., "The Yellow Supergiant Progenitor of the Type II Supernova 2011dh in M51," *The Astrophysical Journal Letters* 739, no. 2 (October 2011): L37.

5. A. Gal-Yam and D. C. Leonard, "A Massive Hypergiant Star as the Progenitor of the Supernova SN 2005gl," *Nature* 458, no. 7240 (April 2009): 865–867.

6. C. McCully et al., "A Luminous, Blue Progenitor System for the Type Iax Supernova 2012Z," *Nature* 512, no. 7512 (August 2014): 54–56.

7. Pre-explosion images of iPTF13bvn, a Type Ib supernova, are inconclusive, allowing for both a single massive star progenitor and a binary system. See Y. Cao et al., "Discovery, Progenitor and Early Evolution of a Stripped Envelope Supernova iPTF13bvn," *The Astrophysical Journal Letters* 775, no. 1 (September 2013): L7. Pre-explosion images of the nearby Type Ia SNe 2011fe and 2014J have ruled out most types of red giants as possible binary companions for the exploding white dwarf. See W. Li et al., "Exclusion of a Luminous Red Giant as a Companion Star to the Progenitor of Supernova SN 2011fe," *Nature* 480, no. 7377 (December 2011): 348–350; O. Graur, D. Maoz, and M. M. Shara, "Progenitor Constraints on the Type-Ia Supernova SN2011fe from Pre-explosion

Hubble Space Telescope He II Narrow-Band Observations," *Monthly Notices of the Royal Astronomical Society* 442, no. 1 (July 2014): L28–L32; O. Graur and T. E. Woods, "Progenitor Constraints on the Type Ia Supernova SN 2014J from *Hubble Space Telescope* Hβ and [O III] Observations," *Monthly Notices of the Royal Astronomical Society* 484, no. 1 (March 2019): L79–L84; and P. L. Kelly et al., "Constraints on the Progenitor System of the Type Ia Supernova 2014J from Pre-explosion Hubble Space Telescope Imaging," *The Astrophysical Journal* 790, no. 1 (July 2014): 3.

8. P. L. Kelly and R. P. Kirshner, "Core-Collapse Supernovae and Host Galaxy Stellar Populations," *The Astrophysical Journal* 759, no. 2 (November 2012): 107; O. Graur et al., "LOSS Revisited. II. The Relative Rates of Different Types of Supernovae Vary between Low- and High-Mass Galaxies," *The Astrophysical Journal* 837, no. 2 (March 2017): 121.

9. O. Graur et al., "LOSS Revisited. I. Unraveling Correlations between Supernova Rates and Galaxy Properties, as Measured in a Reanalysis of the Lick Observatory Supernova Search," *The Astrophysical Journal* 837, no. 2 (March 2017): 120; D. Maoz and O. Graur, "Star Formation, Supernovae, Iron, and α: Consistent Cosmic and Galactic Histories," *The Astrophysical Journal* 848, no. 1 (October 2017): 25.

10. J. S. Bloom et al., "A Compact Degenerate Primary-Star Progenitor of SN 2011fe," *The Astrophysical Journal Letters* 744, no. 2 (January 2012): L17.

11. L. Chomiuk et al., "A Deep Search for Prompt Radio Emission from Thermonuclear Supernovae with the Very Large Array," *The Astrophysical Journal* 821, no. 2 (April 2016): 119; R. Margutti et al., "No X-Rays from the Very Nearby Type Ia SN 2014J: Constraints on Its Environment," *The Astrophysical Journal* 790, no. 1 (July 2014): 52.

12. D. Maoz, F. Mannucci, and G. Nelemans, "Observational Clues to the Progenitors of Type Ia Supernovae," *Annual Review of Astronomy and Astrophysics* 52 (August 2014): 107–170.

13. J. Whelan and I. Iben, Jr., "Binaries and Supernovae of Type I," *The Astrophysical Journal* 186 (December 1973): 1007–1014.

14. R. F. Webbink, "Double White Dwarfs as Progenitors of R Coronae Borealis Stars and Type I Supernovae," *The Astrophysical Journal* 277 (February 1984): 355–360; I. Iben, Jr. and A. V. Tutukov, "The Evolution of Low-Mass Close Binaries Influenced by the Radiation of Gravitational Waves and by a Magnetic Stellar Wind," *The Astrophysical Journal* 284 (September 1984): 719–744. The two white dwarfs merge due to loss of energy and angular momentum to the radiation of gravitational waves (chapter 8).

15. Accretion-induced collapse might create a different, hitherto unseen type of hydrogen-free transient with lower luminosities than Type Ia supernovae.

See K. Nomoto and I. Iben, Jr., "Carbon Ignition in a Rapidly Accreting Degenerate Dwarf: A Clue to the Nature of the Merging Process in Close Binaries," *The Astrophysical Journal* 297 (October 1985): 531–537; S. Darbha et al., "Nickel-Rich Outflows Produced by the Accretion-Induced Collapse of White Dwarfs: Light Curves and Spectra," *Monthly Notices of the Royal Astronomical Society* 409, no. 2 (December 2010): 846–854; and K. J. Shen et al., "The Long-Term Evolution of Double White Dwarf Mergers," *The Astrophysical Journal* 748, no. 1 (March 2012): 35.

16. W. D. Arnett, "A Possible Model of Supernovae: Detonation of ^{12}C," *Astrophysics and Space Science* 5, no. 2 (October 1969): 180–212; W. K. Rose, "Neutrino Emission and the Onset of Carbon Burning," *The Astrophysical Journal* 155, no. 2 (February 1969): 491–499.

17. J. J. Eldridge et al., "The Death of Massive Stars—II. Observational Constraints on the Progenitors of Type Ibc Supernovae," *Monthly Notices of the Royal Astronomical Society* 436, no. 1 (October 2013): 774–795.

18. For a review, see E. Berger, "Short-Duration Gamma-Ray Bursts," *Annual Review of Astronomy and Astrophysics* 52 (August 2014): 43–105.

19. J. Donne, *Devotions upon Emergent Occasions* (London: Printed by A. M. for Thomas Jones, 1624), Meditation XVII.

20. Roughly 40% of Sun-like stars have companions, while for the massive stars that end in core collapse the fraction is higher than 80%. D. Raghavan et al., "A Survey of Stellar Families: Multiplicity of Solar-Type Stars," *The Astrophysical Journal Supplement Series* 190, no. 1 (September 2010): 1–42.

21. H. Sana et al., "Binary Interaction Dominates the Evolution of Massive Stars," *Science* 337, no. 6093 (July 2012): 444–446.

22. E. Zapartas et al., "Delay-Time Distribution of Core-Collapse Supernovae with Late Events Resulting from Binary Interaction," *Astronomy & Astrophysics* 601 (May 2017): A29.

23. J. S. W. Claeys et al., "Theoretical Uncertainties of the Type Ia Supernova Rate," *Astronomy & Astrophysics* 563 (March 2014): A83.

24. M. L. Lidov, "The Evolution of Orbits of Artificial Satellites of Planets under the Action of Gravitational Perturbations of External Bodies," *Planetary and Space Science* 9, no. 10 (October 1962): 719–759; Y. Kozai, "Secular Perturbations of Asteroids with High Inclination and Eccentricity," *The Astronomical Journal* 67, no. 9 (November 1962): 591–598.

25. D. Kushnir et al., "Head-On Collisions of White Dwarfs in Triple Systems Could Explain Type Ia Supernovae," *The Astrophysical Journal Letters* 778, no. 2 (December 2013): L37.

26. A fraction of the energy in the shock goes to breaking iron nuclei in the core of the star into the neutrons that make up the neutron star.

27. H. A. Bethe and J. R. Wilson, "Revival of a Stalled Supernova Shock by Neutrino Heating," *The Astrophysical Journal* 295 (August 1985): 14–23.

28. For a review of theoretical progenitor and explosion models, see B. Wang, "Mass-Accreting White Dwarfs and Type Ia Supernovae," *Research in Astronomy and Astrophysics* 18, no. 5 (May 2018): 049.

29. D. M. Scolnic et al., "The Complete Light-Curve Sample of Spectroscopically Confirmed SNe Ia from Pan-STARRS1 and Cosmological Constraints from the Combined Pantheon Sample," *The Astrophysical Journal* 859, no. 2 (June 2018): 101.

30. L. Kelsey et al., "The Effect of Environment on Type Ia Supernovae in the Dark Energy Survey Three-Year Cosmological Sample," *Monthly Notices of the Royal Astronomical Society* 501, no. 4 (March 2021): 4861–4876.

31. P. L. Kelly et al., "Hubble Residuals of Nearby Type Ia Supernovae Are Correlated with Host Galaxy Masses," *The Astrophysical Journal* 715, no. 2 (June 2010): 743–756.

32. For a concise introduction to electron-capture supernovae, see C. L. Doherty et al., "Super-AGB Stars and Their Role as Electron Capture Supernova Progenitors," *Publications of the Astronomical Society of Australia* 34 (November 2017): e056.

33. G. Hosseinzadeh et al., "Short-Lived Circumstellar Interaction in the Low-Luminosity Type IIP SN 2016bkv," *The Astrophysical Journal* 861, no. 1 (July 2018): 63.

34. A. Jerkstrand et al., "Emission Line Models for the Lowest Mass Core-Collapse Supernovae—I. Case Study of a 9 $M_\odot$ One-Dimensional Neutrino-Driven Explosion," *Monthly Notices of the Royal Astronomical Society* 475, no. 1 (March 2018): 277–305.

35. S. E. Woosley, A. Heger, and T. A. Weaver, "The Evolution and Explosion of Massive Stars," *Reviews of Modern Physics* 74, no. 4 (November 2002): 1015–1071.

36. S. M. Adams et al., "The Search for Failed Supernovae with the Large Binocular Telescope: Constraints from 7 yr of Data," *Monthly Notices of the Royal Astronomical Society* 469, no. 2 (August 2017): 1445–1455.

37. A. Gal-Yam, "The Most Luminous Supernovae," *Annual Review of Astronomy and Astrophysics* 57 (August 2019): 305–333.

38. W. Shakespeare, *Hamlet*, ed. B. A. Mowat and P. Werstine (Folger Shakespeare Library, 2016), 1:5:187–188. Accessed from https://shakespeare.folger .edu/shakespeares-works/hamlet/. The flow of the text has been edited slightly by the author.

39. For reviews, see D. Kasen, "Unusual Supernovae and Alternative Power Sources," in *Handbook of Supernovae* (Springer, 2017), 939–965; and Gal-Yam, "The Most Luminous Supernovae," 305–333.

Chapter 8

1. Harwit estimated there were ~200 distinct phenomena, a quarter of which had already been discovered. However, this analysis only took into account information gathered from light. M. Harwit, "The Number of Class A Phenomena Characterizing the Universe," *Quarterly Journal of the Royal Astronomical Society* 16 (December 1975): 378–409.

2. M. Harwit, *In Search of the True Universe: The Tools, Shaping, and Cost of Cosmological Thought* (Cambridge University Press, 2013), 231–232.

3. This effect, called "seeing," can also be mitigated technologically. "Lucky imaging," for example, has the observer take many short exposures but keep only the sharpest ones. Telescopes with "adaptive optics," on the other hand, use real-time observations of a bright "guide" star or an artificial star created with a strong laser beam to gauge the blurring effect of the atmosphere. The telescope's mirror is then warped according to these measurements in order to mitigate the blurring (many contemporary mirrors are made of segments, and each segment can be tilted slightly to change the overall shape of the mirror). This technique, though powerful, is currently limited to the infrared.

4. D. G. York et al., "The Sloan Digital Sky Survey: Technical Summary," *The Astronomical Journal* 120, no. 3 (September 2000): 1579–1587.

5. Ž. Ivezić et al., "LSST: From Science Drivers to Reference Design and Anticipated Data Products," *The Astrophysical Journal* 873, no. 2 (March 2019): 111.

6. G. Narayan et al., "Machine-Learning-Based Brokers for Real-Time Classification of the LSST Alert Stream," *The Astrophysical Journal Supplement Series* 236, no. 1 (May 2018): 9.

7. M. Lacy, "The Karl G. Jansky Very Large Array Sky Survey (VLASS). Science Case and Survey Design," *Publications of the Astronomical Society of the Pacific* 132, no. 1009 (March 2020): 035001.

8. D. R. Lorimer, "A Bright Millisecond Radio Burst of Extragalactic Origin," *Science* 318, no. 5851 (November 2007): 777–780.

9. P. Cigan et al., "High Angular Resolution ALMA Images of Dust and Molecules in the SN 1987A Ejecta," *The Astrophysical Journal* 886, no. 1 (November 2019): 51.

10. B. P. Abbott et al., "Observation of Gravitational Waves from a Binary Black Hole Merger," *Physical Review Letters* 116, no. 6 (February 2016): 061102.

11. A. Einstein, "Über Gravitationswellen," *Sitzungsberichte der Königlich Preußischen Akademie der Wissenschaften* 1 (January 1918): 154–167.

12. K. Hirata et al., "Observation of a Neutrino Burst from the Supernova SN1987A," *Physical Review Letters* 58, no. 14 (April 1987): 1490–1493; R. M. Bionta et al., "Observation of a Neutrino Burst in Coincidence with Supernova 1987A in the Large Magellanic Cloud," *Physical Review Letters* 58, no. 14 (April 1987): 1494–1496; and E. N. Alekseev et al., "Possible Detection of a Neutrino Signal on 23 February 1987 at the Baksan Underground Scintillation Telescope of the Institute of Nuclear Research," *Journal of Experimental and Theoretical Physics Letters* 45 (May 1987): 589–592.

13. K. Kotake and T. Kuroda, "Gravitational Waves from Core-Collapse Supernovae," in *Handbook of Supernovae* (Springer, 2017): 1671–1698.

14. Sources for plate 8: Gravitational waves: B. P. Abbott et al., "GW170817: Observation of Gravitational Waves from a Binary Neutron Star Inspiral," *Physical Review Letters* 119, no. 16 (October 2017): 161101. Ultraviolet: P. A. Evans et al., "Swift and NuSTAR Observations of GW170817: Detection of a Blue Kilonova," *Science* 358, no. 6370 (December 2017): 1565–1570. Optical and infrared: M. M. Kasliwal et al., "Illuminating Gravitational Waves: A Concordant Picture of Photons from a Neutron Star Merger, *Science* 358, no. 6370 (December 2017): 1559–1565. Radio: G. Hallinan et al., "A Radio Counterpart to a Neutron Star Merger," *Science* 358, no 6370 (December 2017): 1579–1583.

15. P. S. Cowperthwaite et al., "The Electromagnetic Counterpart of the Binary Neutron Star Merger LIGO/Virgo GW170817. II. UV, Optical, and Near-Infrared Light Curves and Comparison to Kilonova Models," *The Astrophysical Journal Letters* 848, no. 2 (October 2017): L17; N. R. Tanvir et al., "The Emergence of a Lanthanide-Rich Kilonova Following the Merger of Two Neutron Stars," *The Astrophysical Journal Letters* 848, no. 2 (October 2017): L27.

16. B. F. Schutz, "Determining the Hubble Constant from Gravitational Wave Observations," *Nature* 323, no. 6086 (September 1986): 310–311.

17. https://wis-tns.org

18. http://www.astronomerstelegram.org

19. http://www.rochesterastronomy.org/snimages

20. https://www.zooniverse.org

21. https://www.zooniverse.org/projects/dwright04/supernova-hunters

22. https://boinc.berkeley.edu

23. https://www.nsf.gov/crssprgm/reu

24. O. Graur, "The Harvard Science Research Mentoring Program," eprint arXiv:1809.08078.

25. https://makersandmentors.org; https://www.nyas.org/programs/global-stem-alliance

FURTHER READING

Bloom, Joshua S. *What Are Gamma-Ray Bursts?* Princeton, NJ: Princeton University Press, 2011.

Kirshner, Robert P. *The Extravagant Universe: Exploding Stars, Dark Energy, and the Accelerating Cosmos*. Princeton, NJ: Princeton University Press, 2002.

Kregenow, Julia. *Twinkle Twinkle Little Star, I Know Exactly What You Are*. Naperville, IL: Sourcebooks, Inc., 2018.

Lang, Kenneth R. *The Life and Death of Stars*. Cambridge, UK: Cambridge University Press, 2013.

Singh, Simon. *Big Bang: The Origin of the Universe*. New York: HarperCollins, 2004.

Stephenson, F. Richard, and David A. Green. *Historical Supernovae and Their Remnants*. Oxford, UK: Oxford University Press, 2002.

Wheeler, J. Craig. *Cosmic Catastrophes: Exploding Stars, Black Holes, and Mapping the Universe*. Cambridge, UK: Cambridge University Press, 2007.

INDEX

Note: All observatories that include the word "observatory" in their name are indexed under Observatory. The same is true for telescopes, which are indexed under Telescope. Observatories and telescopes that do not include these words, such as ALMA, appear on their own. Likewise, all supernova types are indexed under Supernova.

Accretion-induced collapse, 198n15

Adaptive optics, 201n3

Ali ibn Ridwan, 8–10

Amateur astronomy, 160–161, 163, 168. *See also* Mentoring

Annales Beneventani, 10

Aristotle, 15

Asteroid, 33, 97, 102, 180

Astrology, 1, 4, 6, 8, 10, 12, 35, 167

Astronomer's Telegram (ATel), 163

Atacama Large Millimeter/ submillimeter Array (ALMA), 156

AT&T Bell Labs, 29

Atmosphere, xv, 23, 91–92, 95, 146, 158, 201n3

Baade, Walter, 22–23, 185n5

Baryonic matter, 82, 175–176

Bell Labs. *See* AT&T Bell Labs

Berkeley Open Infrastructure for Network Computing (BOINC), 164

Bertaud, Charles, 186n10

Beta decay, 189n11. *See also* Radioactive processes: *e*-process

Beyond Ultra-deep Frontier Fields and Legacy Observations (BUFFALO), 36, 187n19

Biao, Sima, 5–6

Big Bang, 82, 175

Binary systems. *See* Star

Black hole, 69, 72, 76, 78–79, 136– 140, 157, 175, 178, 180–181, 192n5, 192n14

Blueshift, 180. *See also* Redshift

Brahe, Tycho, xvi, 12–15, 19

Brown dwarf, 58, 78, 175

Caltech (California Institute of Technology), 22, 24, 186n12

Carpenter, Edwin Francis, 186n10

Chandrasekhar, Subrahmanyan, 146

Chandrasekhar mass, 125–126, 146

Charge-coupled device (CCD), 28– 35, 105

Chavira, Enrique, 186n10

Cherenkov radiation, 92–93, 158

Circumstellar medium (CSM), 191n10

Citizen science, 160, 163, 168

Cluster. *See* Galaxy; Star

Comet, 3–4, 7–8, 11, 13, 33, 184n17, 185n18
Commentary on the Tetrabiblos of Ptolemy, 8
Compton, Arthur, 146
Constellation, 5–6, 8–9, 14, 16, 96, 184n6
Copernicus, Nicolaus, 15
Cosmic rays, xv, 23, 33, 92–93, 95, 158, 168, 176, 178, 192n2, 193n14. *See also* Radioactive processes: *x*-process
Cosmology, 119, 133
  cosmological constant, 110
  cosmological distance ladder, 100–107
  cosmological model, 15, 110, 175, 196n14
  cosmological redshift, 108, 180 (*see also* Redshift)
  cosmological scale, 82
Crab Nebula. *See* Supernova: remnant

Dark energy, xv, 82, 107, 110–111, 156, 168, 176, 181, 195n9
Dark matter, 81, 112–116, 150, 176
Degeneracy pressure. *See* Pressure
De nova stella. See On the New Star
De revolutionibus orbium coelestium. See On the Revolutions of the Heavenly Spheres
Difference imaging. *See* Imaging
Digges, Thomas, 12

Earth, xv–xvii, 9, 15, 23, 27, 52, 66, 77, 91–93, 95–97, 102, 117, 146, 158, 168, 173, 177, 181, 183n6

Einstein, Albert, 110, 116, 157
Einstein cross, 114, 116, plate 7
Einstein ring, 113, 177
Ejecta. *See* Supernova
Electron-capture supernova. *See* Supernova
Electron degeneracy pressure. *See* Pressure
European Space Agency (ESA), 159
Expansion age. *See* Hubble time

Fast radio burst (FRB), 155–156, 176
Fermi, Enrico, 193n14
Fusion, 59–60, 62, 64, 66, 68, 78–79, 84, 86, 135, 190n1–2
  CNO cycle, 192n4
  p-p chain, 192n4

Galactic wind. *See* Wind
Galaxy
  cluster, 24, 112–114, 116, 177
  dwarf, 31, 118–119
  individual
    Andromeda, 22
    M82, plate 6
    M101, 22
    Magellanic Clouds, 118–119
    Milky Way, xii, xviii, 4, 52, 54–55, 71, 74, 81, 90, 92–93, 97, 100, 103, 118–119, 155, 158, 170, 173, 180
    NGC 266, 123
    NGC 4725, 34
    NGC 6946 ("Fireworks"), 52
  recession, 108, 110–112, 180, 196n15
  starburst, 90, plate 6

as supernova host, xiii, xv, 31, 34,
    50–52, 54, 105, 114, 117–119,
    124, plate 1, plate 7, plate 8
Zoo, 163
Galilei, Galileo, 16–17, 19–20
Gamma-ray burst (GRB), 43, 126,
    177
Gans, David, 11–12, 184n17,
    185n18
Gas pressure. *See* Pressure
Gemma, Cornelius, 12
General relativity, xv, 110–111, 113,
    116, 130, 157, 177
Gravitational lensing, 113–114,
    177
Gravity, xv, 58–62, 64, 69, 82, 89–
    90, 108, 110–111, 125, 156,
    175–176, 178, 191n8
Great Observatories. *See*
    Observatory
Guest star, 5–6, 8, 167

Hagecius, Thaddeus, 13
Haro, Guillermo, 186n10
Harwit, Martin, 143–144, 146,
    201n1
Herodianus (Herodian), 6
Hess, Victor, 23, 91–92
Historia Augusta, 7
Historical records of supernovae.
    See Supernova: historical
    records
History of the Later Han Dynasty
    (*Hou Hanshu*), 5
Host galaxy. *See* Galaxy
Hubble, Edwin, 110–111, 146
Hubble constant (H_0), 111–112, 114,
    190n15, 196n15
Hubble flow, 196n14

Hubble's law, 111–112, 196n15
Hubble time, 112

Imaging, 154, 192n14
    difference, 32–36, 154, plate 2
    lucky, 201n3
Intergalactic medium (IGM),
    193n10
Intermediate-mass elements, 131–
    132
International Astronomical Union
    (IAU), 24, 161
Interstellar medium (ISM), 191n10,
    192n13
Inverse beta decay. *See* Beta decay
Iron-group elements, 85–87, 131–
    132

Jansky, Karl Guthe, 155
Jonson, Josef J., 24, 34
Jupiter, 13, 17, 175

Karl G. Jansky Very Large Array
    (VLA), 155
Kepler, Johannes, 15–16
Kilonova, 159–160, 177
    GW170817, 149, 159, plate 8
Kirshner, Robert P., 186n12
Kozai mechanism. *See* Lidov-Kozai
    mechanism
Kukarkin, Boris Vasilyevich, 186n10

Laser Interferometer Space Antenna
    (LISA), 159
Leavitt, Henrietta Swan, 103, 105,
    195n3
Legacy Survey of Space and Time
    (LSST), 143, 149–155, 168
Lensing. *See* Gravitational lensing

Lick Observatory Supernova Search
(LOSS), 32
Lidov-Kozai mechanism, 129, 195n6
Light curve, 42–45, 47–50, 66, 71–
72, 105–109, 129–130, 133,
140, 177, 181
Light echo, 117–118, 177
Luminosity distance, 107–108,
190n15
Lundmark, Knut, 22, 185n2

Macronova, 159, 177. *See also*
Kilonova
Magnetar, 139–140, 178, 180
Main sequence, 59, 62–63, 78,
178–179
Mars, xii, 8, 13, 102
Meitner, Lise, 195n6
Meitner-Auger effect, 195n6
Mentoring, 165
Messenger, 158–159. *See also*
Multimessenger astronomy
Metal, 82, 84, 95, 119
Metallicity, 87
Meteor, 3–4, 13
Milky Way Galaxy. *See* Galaxy
Minkowski, Rudolph, 42
Molecular cloud, 58, 60, 78, 89–91,
178
Moon, xvii, 3, 8–9, 13, 17, 20, 93,
95, 102
Moons, xvii, 21
Multimessenger astronomy, 143,
157–160, 168, 178
Muñoz, Jerome, 12

NASA (National Aeronautics and
Space Administration), 146,
150, 156

National Science Foundation (NSF),
164
Nebula, 72, 178–180
absorption and emission, 178
extragalactic, 24
nebular phase, 50
planetary, 66, 179, 191n6
pulsar-wind, 77, 79, 179
Neutron degeneracy pressure. *See*
Pressure
Neutronization, 69, 191n8
Neutron star, 22–23, 69, 71–72,
75–79, 86, 125, 130–131,
135–136, 139, 146, 157, 159,
176–179, 181, 191n5, 192n5,
200n26, plate 8
Nova, xi, 3, 19, 21–23, 38, 149, 178,
180, 185n2
Nova stella. *See* Stella nova
Nucleosynthesis, 84, 86, 178, 192n6

Observatory, 24–26, 99, 143–144,
146–148, 150, 158–159
Asiago Astrophysical Observatory,
186n10
Chandra X-ray Observatory, 146
Compton Gamma Ray
Observatory, 146
Córdoba Observatory, 186n10
Great Observatories, 146–147
Laser Interferometer
Gravitational-Wave
Observatory (LIGO), 157,
plate 8
Lick Observatory, 32
Meudon Observatory, 186n10
Moscow University Observatory,
186n10
Mount Wilson Observatory, 22

Palomar Observatory, 24–26
Rubin Observatory, 143, 150–152
Steward Observatory, 186n10
Tonantzintla Observatory, 186n10
Yerkes Observatory, 20
Zimmerwald Observatory, 186n10
On the New Star, 13
On the Revolutions of the Heavenly Spheres, 15
Outflow, 90–91, 176, 179, plate 6. *See also* Wind

Pair-instability supernovae. *See* Supernova
Pan-STARRS1, 164
Pauli exclusion principle, 191n5
Periodic table, xv, 58, 82, 84–85, plate 4
Phase space, 143–149
Phillips, Mark, 105
Phillips relation. *See* Rust-Pskovskii-Phillips relation
Photodisintegration, 86, 191n8
Planet, xi, xvii–xviii, 3–4, 8–9, 13, 20, 81, 93, 100, 102, 168, 175, 191n6
Planetary nebula. *See* Nebula
Pliocene-Pleistocene boundary, xvi, 95
Pressure, 58, 62, 64, 69, 125, 135, 175–176, 191n5, 191n8
  degeneracy, 64, 176, 191n5
  electron, 64, 125–126, 176, 181, 191n8
  neutron, 69, 176, 178
  gas, 58–62, 72, 126, 136, 190n1

gravitational, xv, 58–60, 69, 126, 176, 178, 191n5
radiation, 59, 190n1
Progenitor, xiii, 97, 121–124, 126–127, 134, 137, 159, 179, 197n7, 200n28
Pskovskii, Yuri Pavlovich, 105, 195n6
Ptolemaeus, Claudius (Ptolemy), 8, 186n16
Pulsar, 76–77, 179
Pulsar-wind nebula. *See* Nebula

Quantum degeneracy pressure. *See* Pressure

Radiation pressure. *See* Pressure
Radioactive processes, 48, 50, 71, 140, 158–159, 174, 189n11
  α-process, 84
  e-process, 189n11 (*see also* Beta decay)
  r-process, 85–86, 177
  s-process, 85–86
  x-process, 192n2 (*see also* Cosmic rays)
Rapidly evolving transient. *See* Transient
Red giant, 59, 64, 68, 78, 84, 99, 125, 179, 197n7
Redshift, 108–109, 112, 119, 156, 160, 180, 190n15, 196n14–15. *See also* Blueshift
Relativity. *See* General relativity
Remnant. *See* Star: stellar remnant; Supernova: remnant
Roman, Nancy Grace, 156
Romano, Giuliano, 186n10
Rosino, Leonida, 186n10

Rubin, Vera Cooper, 112–113
Rust, Bert Woodward, 105, 195n6
Rust-Pskovskii-Phillips relation, 105, 195n6

Saturn, 13, 17
Schürer, Max, 186n10
Sefer Zemah David. See Zemah David
Sérsic, José Luis, 186n10
Sherlock Holmes, xi, 161. *See also* Supernova: in popular culture
Shock, 47–49, 69, 72, 74–76, 89–90, 92, 117, 130–132, 136, 191n9, 192n12, 193n14, 200n26
  breakout, 47, 188n8
  forward, 72, 74–76, 92, 193n8
  reverse, 74–75
Sloan Digital Sky Survey (SDSS), 152, 163
Solar System, xvi–xvii, 93, 97, 150, 168, 175, 194n19
Space-time, 113–114, 157, 177
Spallation. *See* Radioactive processes: *x*-process
Spectrograph, 42
Spectroscopy, xviii, 154
Spectrum, 42, 152, 180
  of cosmic-ray energy, 95
  electromagnetic, 40–41, 71, 155, 160, 180, 189n11
  spectral resolution, 144
  of supernova, 42–43, 45–46, 49–50, 66, 108–109, 117, 126, 129, 131, 138–139, 161
  synthetic, 130
Spitzer, Lyman, 146
Standard candle, 103–105, 114, 132–133, 160
Standard siren, 160

Star
  in binary systems, xiii, 96, 122, 124, 126–129, 159, 197n7
  cluster, 21, 89–90, 147
  individual
    Betelgeuse, 96
    Eta Carinae, 96
    IK Pegasi, 96
    Vega, 29
  massive, 72, 86, 96, 122, 126–127, 176, 180, 190n1, 192n6, 197n7, 199n20, plate 5, plate 6
  Population III, 137
  stellar remnant, 22, 76, 79, 137, 178, 181
  supermassive, 138
  in triple systems, 127–129
  very massive, 137, 179–180
Stella nova, xvii, 13–14, 16, 19
Strong lensing. *See* Gravitational lensing
Sun, xi, xiii, xvi–xviii, 8–9, 15, 29, 62–63, 65–66, 72, 76, 87, 91–93, 96–97, 102, 127, 173, 190n2, 199n20
Superbubble, 89–90, 180, plate 5
Supergiant, 59, 63, 68, 78, 96, 122
Supermassive black hole. *See* Black hole
Supernova
  calcium-rich, 43, 149
  classification, 41–47
  core-collapse, 43, 66, 69, 74, 76, 78, 86, 96, 124–125, 127, 129–130, 138, 149, 159, 175, 180, 189n15, 192n14
  ejecta, 47–50, 72, 74–75, 117, 139–140, 176, 180, 189n11, 191n9, 192n12

electron-capture, 78–79, 138, 176, 200n32

etymology, xvi–xvii, 21–23, 185n2

failed, 136

historical records, 3–17

individual

  BUF19Awd, v, 35–36, 187n19, plate 2

  BUF19Sop, v, 35–36, 187n19, plate 2

  iPTF13bvn, 197n7

  SN 185, 4–7

  SN 386, 4

  SN 393, 4

  SN 1006, 4, 7–10, plate 3

  SN 1054, 4, 79, plate 3

  SN 1181, 4, 79, plate 3

  SN 1572, 4, 10–15, 22, 74, plate 3

  SN 1604, 4, 10, 15–16, 22, 52, plate 3

  SN 1885A (S Andromeda), 22

  SN 1909A (SS Uma), 22

  SN 1940B, 34

  SN 1987A, xvii, 44, 71, 122, 158

  SN 2005gl, 123

  SN 2009ip, 44

  SN 2011fe, 197n7

  SN 2012Z, 122

  SN 2014J, 197n7

  SN 2016bkv, 135

  SN 2018gv, plate 1

  SN Refsdal, 116, plate 7

and mass extinction, xvi, 81, 93–97

pair-instability, 71, 78–79, 136–138, 179, 192n14

in popular culture, xi–xii

rates, 50–55

remnant, 4, 10, 13, 55, 71–72, 74–76, 87, 89–92, 117, 156, 176, 180, 192n12, plate 3

  3C 58, 74, 79, plate 3

  Cassiopeia A, 55, 75

  Crab Nebula, 74, 77, 79, plate 3

  G1.9+0.3, 55

  G327.6+14.6, 10, plate 3

  Kepler, 15, plate 3

  Tycho (G120.1+1.4), 13, 74, plate 3

stripped-envelope, 43, 175

superluminous, 43, 107, 137–140, 149, 178, 180

survey, 23–28, 31–32, 34, 36, 38, 41, 52, 133, 143, 150–156, 160–161, 164, 168, 186n10, 186n12, 187n19

Type I, 42, 48

Type Ia, 43–44, 46, 50, 52, 64, 66–67, 74, 78, 86, 96, 100, 105, 107, 112, 114, 116, 124–125, 127, 129, 131, 133–134, 149, 156, 159–160, 181, 188n7–8, 189n15, 197n7, 198n15

Type Ia-91bg, 43, 188n7

Type Ia-91T, 43, 188n7

Type Ia-Normal, 43, 188n7

Type Ian, 43

Type Iax, 43, 122, 149

Type Ib, 43–44, 46, 69–70, 126, 197n7

Type Ic, 43–44, 46, 69–70, 124, 126

Type Ic-BL, 43, 126, 177

Type Ib/c, 43, 50, 52, 124, 127, 175, 188n8

Type II, 42–43, 48, 50, 52, 69–70, 175, 188n8

Supernova (cont.)
  Type II-87A, 43–44
  Type IIb, 43, 48, 75, 122, 126
  Type IIL, 43–44, 48, 188n7
  Type IIn, 43–44, 46, 49–50, 72, 122
  Type IIP, 43–44, 46, 48, 50, 107, 122, 135, 188n7
  Types III, IV, and V, 187n3

Telescope, 17, 19–21, 24–28, 34–36, 39, 41, 71, 74, 77, 93, 99, 117, 138, 143–144, 146–149, 152, 154–156, 158, 161, 163, 168, 185n1, 186n11, 190n15, 191n6, 192n14, 201n3
  extremely large telescopes (ELTs), 138, 148–150
  individual
    European Extremely Large Telescope, 150
    Giant Magellan Telescope, 150
    Hubble Space Telescope (HST), 27, 35–36, 112, 114, 143, 146, 150, 156, plate 2, plate 8
    James Webb Space Telescope (JWST), 150, 156
    Roman Space Telescope, 156
    Spitzer Space Telescope, 146
    Thirty Meter Telescope, 150
  refracting and reflecting, 19–20
Tidal disruption event (TDE), 180
Time-domain astronomy, 143, 168, 180
Transient, 25–26, 140, 148–149, 152–156, 159, 161, 163, 168, 176–181, 198n15
  rapidly evolving, 43, 140, 149, 154, 178–179

Transient Name Server (TNS), 161, 163
Triple system. See Star
Tycho. See Brahe, Tycho; Supernova: remnant

Venus, 8, 17
Very Large Array. See Karl G. Jansky Very Large Array (VLA)

Weak lensing. See Gravitational lensing
Webb, James, 150
White dwarf, 63, 65–67, 76, 78, 86–87, 96, 99, 122, 124–127, 131–132, 135, 146, 159, 176–179, 181, 191n5, 197n7, 198n14
Wild, Paul, 186n10
Wind, 139–140
  galactic, 89–90, 176, 179
  pulsar-wind nebula (see Nebula)
  stellar, 49, 66, 72, 87, 89–90, 117, 126, 137–138, 176, 180, plate 5, plate 6

Zel'dovich, Yakov Borisovich, 110
Zemah David, 11
Zodiac. See Constellation
Zooniverse, 163
Zwicky, Fritz, 22–28, 31, 34–35, 41, 112–113, 185n5, 186n12, 187n3
Zwicky Transient Facility (ZTF), 25–26

The MIT Press Essential Knowledge Series

AI Assistants, Roberto Pieraccini
AI Ethics, Mark Coeckelbergh
Algorithms, Panos Louridas
Annotation, Remi Kalir and Antero Garcia
Anticorruption, Robert I. Rotberg
Auctions, Timothy P. Hubbard and Harry J. Paarsch
Behavioral Insights, Michael Hallsworth and Elspeth Kirkman
Biofabrication, Ritu Raman
The Book, Amaranth Borsuk
Carbon Capture, Howard J. Herzog
Citizenship, Dimitry Kochenov
Cloud Computing, Nayan B. Ruparelia
Collaborative Society, Dariusz Jemielniak and Aleksandra Przegalinska
Computational Thinking, Peter J. Denning and Matti Tedre
Computing: A Concise History, Paul E. Ceruzzi
The Conscious Mind, Zoltan E. Torey
Contraception, Donna J. Drucker
Critical Thinking, Jonathan Haber
Crowdsourcing, Daren C. Brabham
Cynicism, Ansgar Allen
Data Science, John D. Kelleher and Brendan Tierney
Death and Dying, Nicole Piemonte and Shawn Abreu
Deconstruction, David J. Gunkel
Deep Learning, John Kelleher
Extraterrestrials, Wade Roush
Extremism, J. M. Berger
Fake Photos, Hany Farid
fMRI, Peter A. Bandettini
Food, Fabio Parasecoli
Free Will, Mark Balaguer
The Future, Nick Montfort
Gender(s), Kathryn Bond Stockton
GPS, Paul E. Ceruzzi
Haptics, Lynette A. Jones
Hate Speech, Caitlin Ring Carlson
Information and Society, Michael Buckland
Information and the Modern Corporation, James W. Cortada

Intellectual Property Strategy, John Palfrey
The Internet of Things, Samuel Greengard
Irony and Sarcasm, Roger Kreuz
Ketamine, Bita Moghaddam
Machine Learning: The New AI, Ethem Alpaydın
Machine Translation, Thierry Poibeau
Macroeconomics, Felipe Larraín B.
Memes in Digital Culture, Limor Shifman
Metadata, Jeffrey Pomerantz
The Mind–Body Problem, Jonathan Westphal
MOOCs, Jonathan Haber
Neuroplasticity, Moheb Costandi
Nihilism, Nolen Gertz
Open Access, Peter Suber
Paradox, Margaret Cuonzo
Phenomenology, Chad Engelland
Post-Truth, Lee McIntyre
Quantum Entanglement, Jed Brody
Recommendation Engines, Michael Schrage
Recycling, Finn Arne Jorgensen
Robots, John Jordan
School Choice, David R. Garcia
Science Fiction, Sherryl Vint
Self-Tracking, Gina Neff and Dawn Nafus
Sexual Consent, Milena Popova
Smart Cities, Germaine R. Halegoua
Spaceflight, Michael J. Neufeld
Spatial Computing, Shashi Shekhar and Pamela Vold
Supernova, Or Graur
Sustainability, Kent E. Portney
Synesthesia, Richard E. Cytowic
The Technological Singularity, Murray Shanahan
3D Printing, John Jordan
Understanding Beliefs, Nils J. Nilsson
Virtual Reality, Samuel Greengard
Visual Culture, Alexis L. Boylan
Waves, Frederic Raichlen

DR. OR GRAUR is a Senior Lecturer in Astrophysics at the Institute of Cosmology and Gravitation at the University of Portsmouth and a Research Associate at the American Museum of Natural History. He studies the progenitors of supernovae and other transient phenomena using ground- and space-based telescopes.